Star Origami

AK Peters/CRC Recreational Mathematics Series

The Untold Story of Everything Digital
Bright Boys, Revisited
Tom Green

Wondrous One Sheet Origami
Meenakshi Mukerji

The Geometry of Musical Rhythm
What Makes a "Good" Rhythm Good?, Second Edition
Godfried T. Toussaint

A New Year's Present from a Mathematician
Snezana Lawrence

Six Simple Twists
The Pleat Pattern Approach to Origami Tessellation Design, Second Edition
Benjamin DiLeonardo-Parker

Tessellations
Mathematics, Art, and Recreation
Robert Fathauer

Mathematics of Casino Carnival Games
Mark Bollman

Mathematical Puzzles
Peter Winkler

X Marks the Spot
The Lost Inheritance of Mathematics
Richard Garfinkle, David Garfinkle

Luck, Logic, and White Lies
The Mathematics of Games, Second Edition
Jörg Bewersdorff

Mathematics of The Big Four Casino Table Games
Blackjack, Baccarat, Craps, & Roulette
Mark Bollman

Star Origami
The Starrygami™ Galaxy of Modular Origami Stars, Rings and Wreaths
Tung Ken Lam

For more information about this series please visit: https://www.routledge.com/AK-PetersCRC-Recreational-Mathematics-Series/book-series/RECMATH?pd=published,forthcoming&pg=2&pp=12&so=pub&view=list

Star Origami

The Starrygami™ Galaxy of Modular Origami Stars, Rings and Wreaths

Tung Ken Lam

CRC Press is an imprint of the
Taylor & Francis Group, an **informa** business

First edition published 2022
by CRC Press
6000 Broken Sound Parkway NW, Suite 300, Boca Raton, FL 33487-2742

and by CRC Press
2 Park Square, Milton Park, Abingdon, Oxon, OX14 4RN

CRC Press is an imprint of Taylor & Francis Group, LLC

Library of Congress Cataloging-in-Publication Data

Names: Lam, Tung Ken, author.
Title: Star origami : The Starrygami, galaxy of modular origami stars,
rings and wreaths / Tung Ken Lam.
Description: First edition. | Boca Raton : CRC Press, 2021. | Includes
bibliographical references and index.
Identifiers: LCCN 2021019448 (print) | LCCN 2021019449 (ebook) | ISBN
9781032026626 (hardback) | ISBN 9781032022338 (paperback) | ISBN
9781003184492 (ebook)
Subjects: LCSH: Origami--Mathematics. | Origami in education. | Star
(Shape) | Geometry--Study and teaching--Audio-visual aids. |
Origami--Design.
Classification: LCC TT870 .L245 2021 (print) | LCC TT870 (ebook) | DDC
736/.982--dc23
LC record available at https://lccn.loc.gov/2021019448
LC ebook record available at https://lccn.loc.gov/2021019449

ISBN: 978-1-032-02662-6 (hbk)
ISBN: 978-1-032-02233-8 (pbk)
ISBN: 978-1-003-18449-2 (ebk)

DOI: 10.1201/9781003184492

Publisher's note: This book has been prepared from camera-ready copy provided by the author.

Contents

Ring Stars

Wreaths

Stars

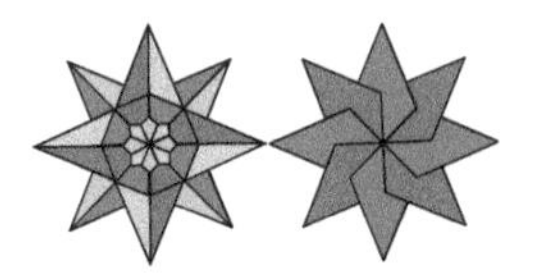

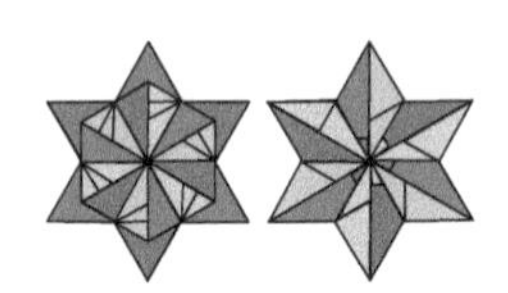

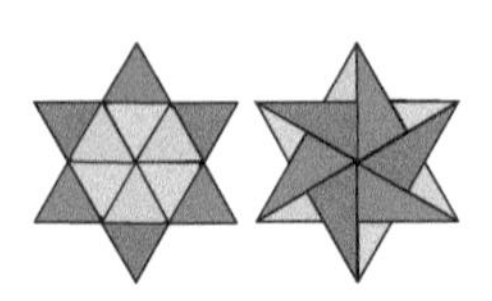

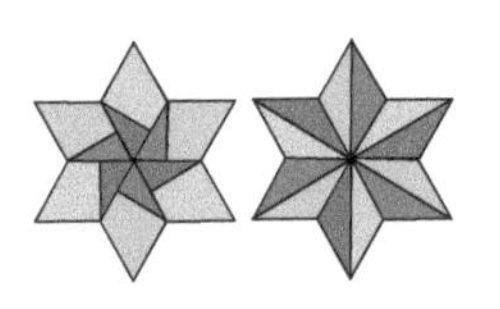

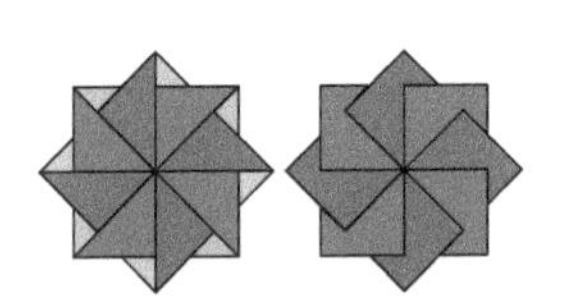

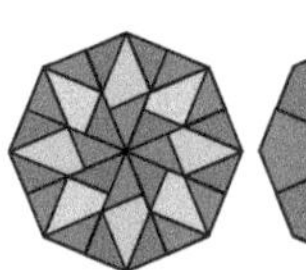

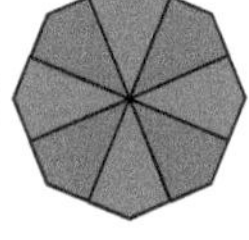

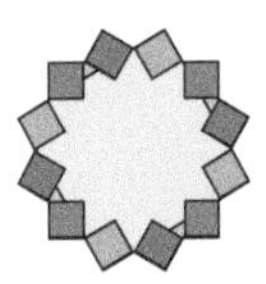

Sliders

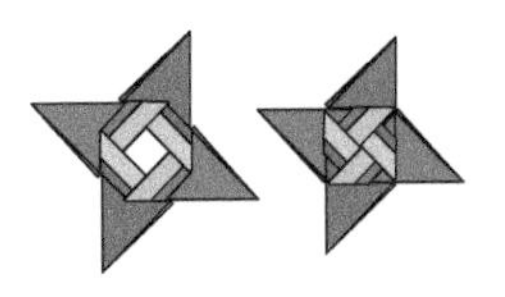

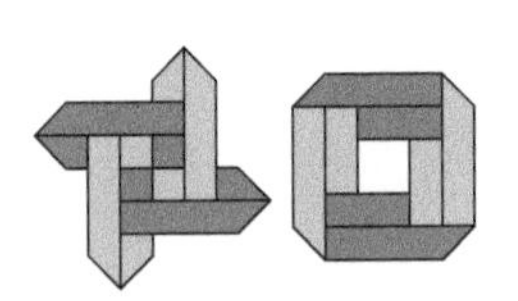

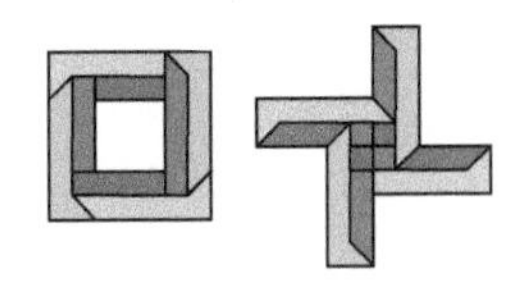

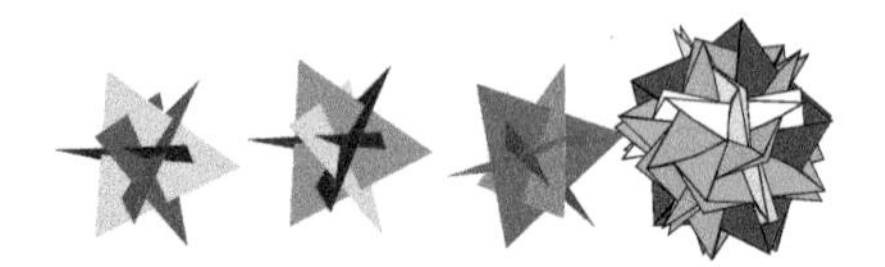

Chapter 1

Introduction

This book is a collection of origami rings, stars and wreaths made using the modular technique. You fold a number of simple units that you join together — without glue — like a puzzle. Some models have bonus features: sliders change shape when you slide the pieces. Others show different patterns when held up to a strong light. Some models have different front and back faces.

1.1 Stars

What is a star? It is a bright fixed point in the night sky, or its stylised representation. For the purposes of this book, a star is an object with pointy arms radiating from its centre. Rings have a hole in the centre. Pinwheels and wreaths have rotational symmetry without reflective symmetry. Kaleidoscope stars have reflective and rotational symmetry. However, these are loose definitions. The chapter entitled Stars, p. 5, gives some ideas on how to visualise and construct stars.

Stars are popular subjects in origami, both using a single sheet and the modular method. Unlike representational origami like animals and birds, stars are a "pure" kind of origami unaffected by "realism". As well as appearance, some criteria for the quality of stars are the elegance of the folding sequence, efficient use of paper and the effectiveness of the joining method.

1.2 Origami stars and how this book is organised

Most of the origami projects in this book are straightforward and quick to make. They use rectangles that have proportions that make effective and efficient units: these rectangles are described in Special Rectangles, p. 15. The chapter Tips on Paper and Folding, p. 37, gives some advice on choosing and preparing paper. It also suggests how to fold and join the units. This book uses standard origami symbols and terminology found in most modern books. They are described in the chapter titled Symbols and Procedures, p. 41.

As you work your way through the projects (starting on p. 45), their structures and patterns will become apparent. Origami stars are a practical and gentle

DOI: 10.1201/9781003184492-1

way of exploring geometry and mathematics. There is a special chapter of advice and ideas for teachers (Notes for Teachers, p. 112). Once you have made a model, it's natural to ask if it can be varied and generalised. For example, if a star uses eight units, is it possible to vary the unit and join a different number of units? These kinds of questions are explored in Folding and Generalising Stars, p. 27.

1.3 How to use this book

1.3.1 For beginning folders

You can browse the book for a simple model to start with. Take time to read the instructions and study each step and the following step to see the effect of each fold. If needed, check Symbols and Procedures, p. 41, to understand the symbols: the key thing to understand is the difference between valley and mountain folds.

Read Tips on Paper and Folding, p. 37, for some helpful hints. You may find Common folding problems and how to avoid them, p. 112, useful for another perspective on successful folding.

1.3.2 For intermediate and experienced folders

You are welcome to dive in and start folding anything that you like the look of. However, please consider the suggestions in the next two subsections.

1.3.3 For potential and experienced creators

Folding and Generalising Stars, p. 27, shows you how to vary existing stars and create new ones. Alternatively, pick a model and analyse it. Special Rectangles, p. 15, might help you. As Robert Lang wrote,

> ability comes with practice...The budding origami designer develops his or her ability by designing and seeing the result. Design can start simply by modifying an existing fold. Make a change; see the result. ...The great leap ...arises from the development of an understanding of why: Why did the designer do it that way?
>
> ...To learn to design, you must disregard reverence for another's model, and be willing to pull it apart, fold it differently, change it and see the effects of your changes. [Lang 11, p. 6]

Experienced creators may wish to explore the possibilities of special rectangles (figure 3.3, 17)). Novel work is likely to come from using geometries that are less frequently explored, e.g. 60°.

1.3.4 To use specific paper or make a particular subject

Use the index to find models by starting shape or number of units. For example, you might only have two sheets of A4 in colours that combine well. A4 is a silver rectangle (figure 3.3, 17). What could you make? Scanning the index for "silver" reveals more than a dozen models. Each sheet of A4 could be cut into two A5 or four A6 rectangles. It could also be cut into three bronze rectangles. So you could choose a model with 4, 6 or 8 units. Models with odd numbers like 5 and 7 may work better in a single colour, so each sheet could be cut into eight A7 rectangles. If you make two five-piece models, the remaining rectangles could make a six-piece model.

Looking again at the index, there are more possibilities as there are entries for "half silver" and "third silver". Another approach is to cut each A4 sheet into a square and a leftover rectangle, then cut each rectangle into four. You now have eight squares and eight leftover rectangles which gives you even more options.

Another example: you want to make a six-pointed star. Looking up "6" in the index shows more than a dozen models, including the 3D *XYZ Rhombic* (p. 106). Alternatively, browse the visual table of contents for an appealing six-pointed star of suitable difficulty.

1.3.5 For teachers

Origami is a practical and gentle way of exploring geometry and mathematics. Notes for Teachers, p. 112, gives advice and ideas for using origami for teaching and learning mathematics. Origami and mathematics education, p. 130, describes some other books for more ideas.

The material in Stars, p. 5, can be implemented in media other than origami. You could use non-digital media like pen/pencil and paper, circular geoboards and curve stitching. Some digital methods include vector drawing programs, virtual geoboards (https://nrich.maths.org/2883), the Logo programming language and dynamic geometry software like Geogebra.

1.4 The models

The projects are grouped by appearance and technique:

Ring stars (p. 45) are stars with a hole in the middle.

Wreaths (p. 59) also have a hole in the middle. They have eight or more points that are right angles or obtuse.

Stars (p. 72) have acute points and have a negligible central hole. They can be harder to make than models with a larger central hole as they are usually less forgiving of inaccuracy than ring stars and wreaths.

Sliders (p. 92) change shape when you slide the units.

3D stars (p. 104) do not lie flat. These could fill an entire book, so the selected 3D stars are planar, i.e. consist of intersecting polygons. They do not have a conventional "inside" like other kinds of 3D stars like the small stellated dodecahedron.

It is sometimes hard to name origami models, especially when they are geometric shapes without a common name. Some creators use numbers, names of people, stars, mountains or other objects. The names can be evocative but they do not suggest what the models looks like. I tried to choose model names that combine the final appearance, the starting shape, the number of units (or number of points of the star), the principal fold and joining method. For example, *Star 6 Bronze Wrap* (p. 82) tells you that it is a six-pointed star made from bronze rectangles and that the units are wrapped together in some way. On the other hand, some names like *WXYZ* (p. 109) are meaningful when you understand the structure.

1.5 Difficulty

Separate difficulty ratings are given for folding the units and then for assembly.

1.5.1 Difficulty of folding the units

Most units are simple to intermediate:

Simple * The only folds are valley and mountain folds with no more than about a dozen steps, e.g. *Star 4, Square From Silver Rectangle* (p. 46) and *Star 4 Leftover Wrap* (p. 81).

Low Intermediate ** Only a few compound folds like reverse folds and squash folds, e.g. *Octagram Paper Cup - Kite Wheel* (p. 63).

Intermediate *** Several compound folds like reverse folds and squash folds, e.g. *Star 6 Fancy Wrap* (p. 86).

High Intermediate **** May have a few tricky folds like multiple sinks or need careful adjustment for well-fitting units.

Complex ***** High accuracy is essential. May have several difficult folds or need extensive precreasing.

1.5.2 Difficulty of assembly

Simple * Each unit attaches easily to the next and does not come apart easily, e.g. *Star 4, Square From Silver Rectangle* (p. 46).

Low Intermediate ** Each unit attaches easily to the next but is only partially stable *Octagram From LOR Wrap Inside* (p. 87).

Intermediate *** Each unit attaches fairly easily to the next but is not particularly stable until several units are combined, e.g. *Square Pointer Slider* (p. 94), or attaching each unit is not straightforward, or the last unit is difficult to attach, e.g. *WXYZ* (p. 109).

High Intermediate **** Careful attention is needed for the order of joining, or units can be hard to join, e.g. *Blintz Icosidodecahedron* (p. 111).

Complex ***** High accuracy is essential. Units do not securely attach to the next, or are awkward to join.

Chapter 2

Stars

There are many ways of making two-dimensional stars. This chapter describes a few methods that can be implemented by drawing, computer graphics, dynamic geometry, origami, etc. First, here is a reminder, if you need it, about the definition and properties of regular polygons.

2.1 Regular polygons

A *polygon* is a two-dimensional shape with straight edges (sides). Polygons can be *convex* or *concave*. A simple test is to place a line on each edge, one at a time. If at least one line crosses at least one other edge of the polygon then it is concave, otherwise it is convex.

A *regular polygon* is a polygon whose sides (edges) are all equal and whose interior angles are all equal (figure 2.1). The *interior angle* is the angle between edges at a vertex (corner). The *central angle* is the angle subtended by an edge at the centre: it has the same value as the *exterior angle* which is $180° - interior\ angle$. Table 2.1 shows the interior and exterior angles for n between three and 20. It also shows the name.

DOI: 10.1201/9781003184492-2

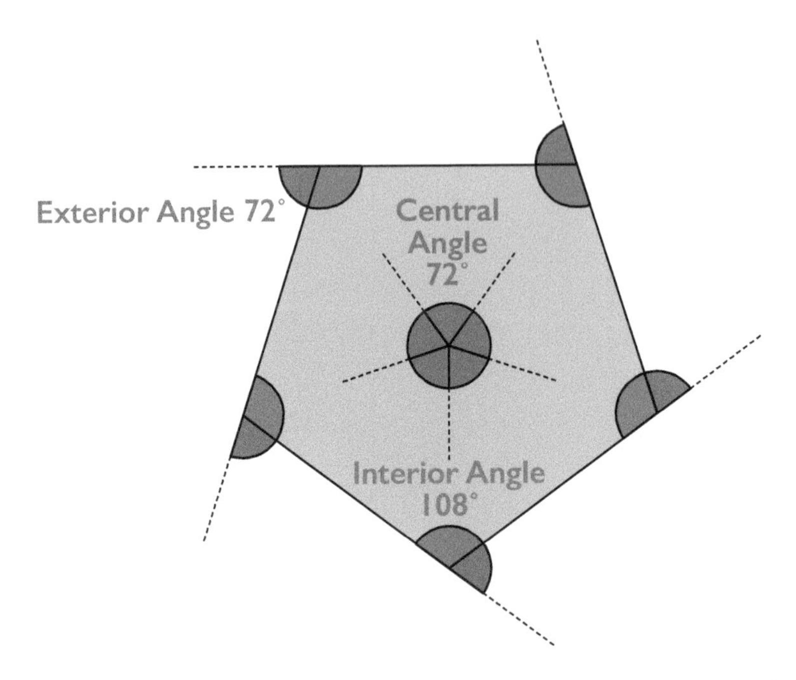

Figure 2.1: The central and exterior angles are the same in a regular polygon like this pentagon. The central angle is 360° divided by the number of sides. The sum of the exterior and interior angles is 180°.

You can drawn polygons using turtle graphics on a computer. For example, define a procedure in Logo that takes two arguments, the length of an edge and the number of sides:

```
TO POLYGON :length :sides
REPEAT :sides [
FD :length
RT 360/:sides
]
END

HOME
CS
POLYGON 100 5
```

The turtle is sent to its home position and has bearing 0°, i.e. facing north, heading up the screen/page. The turtle moves forward 100 units, drawing a straight line, and then turns right by 360°÷5, i.e. 72°. It does this four more times (for a total of five times) to arrive back at the starting point, facing the same direction (same heading or bearing). The turtle has drawn a regular pentagon. In some implementations of Logo you can call the function FILL to fill the polygon with the current FILLCOLOR.

The rest of this chapter describes some methods for making two-dimensional stars.

Table 2.1: Names of selected regular polygons with exterior and interior angles given up to 3 decimal places.

n	*Regular polygon name of* n-gon	*exterior angle*°	*interior angle*°
3	equilateral triangle	120	60
4	square	90	90
5	pentagon	72	108
6	hexagon	60	120
7	heptagon (or septagon)	51.429	128.571
8	octagon	45	135
9	nonagon (or enneagon)	40	140
10	decagon	36	144
11	hendecagon (or undecagon)	32.727	147.273
12	dodecagon	30	150
13	triskaidecagon	27.692	152.308
14	tetrakaidecagon	25.714	154.286
15	pentadecagon	24	156
16	hexakaidecagon	22.5	157.5
17	heptadecagon	21.176	158.824
18	octakaidecagon	20	160
19	enneadecagon	18.947	161.053
20	icosagon	18	162

2.2 Overlapping regular polygons

Take two congruent squares and place one on top of the other. Rotate one square to make an eight-pointed star: the centre of rotation is the centre of the squares and the angle of rotation is half of the central angle of the square, i.e. 45°. This star is an octagram (figure 2.2) and is known as the Star of Lakshmi. It also forms the Islamic symbol Rub el Hizb.

A similar process applied to an equilateral triangle makes a hexagram which is also known as the Star of David. Applying the process to regular polygons with more sides can make the result harder to distinguish from circles as the interior angle approaches 180° (figure 2.5).

For a dodecagon, an alternative approach is to overlap three squares or four equilateral triangles. A 16-gon can have four overlapping squares. (figure 2.5, bottom right).

2.3 Stellating a regular polygon

Stellating a polygon means extending its edges so that they meet. For example, take a regular octagon and extend the sides. End the lines where they meet other lines. This makes an octagram (figure 2.3, left). However, if you extend the lines further, there is another end point that you can choose (figure 2.4, right).

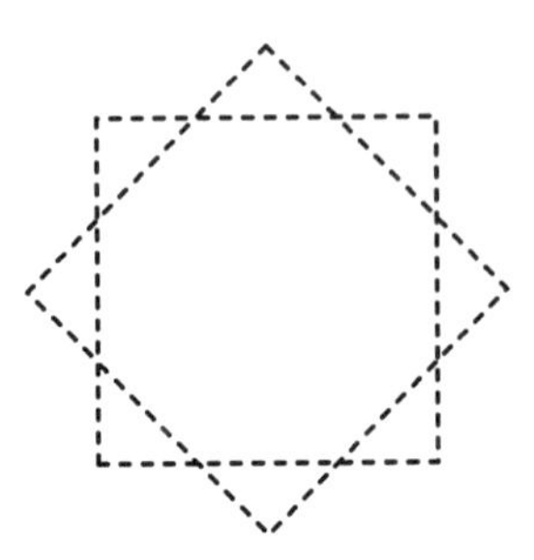

Figure 2.2: Overlap two squares and rotate one about the centre by half of the central angle.

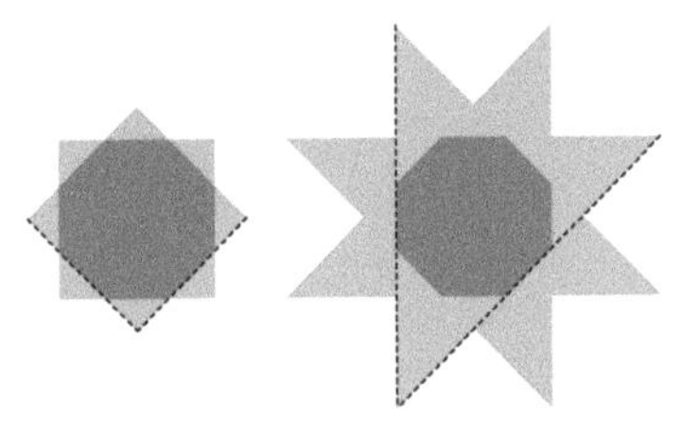

Figure 2.3: Two octagrams made by stellating a regular octagon. The extended edges of the octagon can meet close (left) or further away (right).

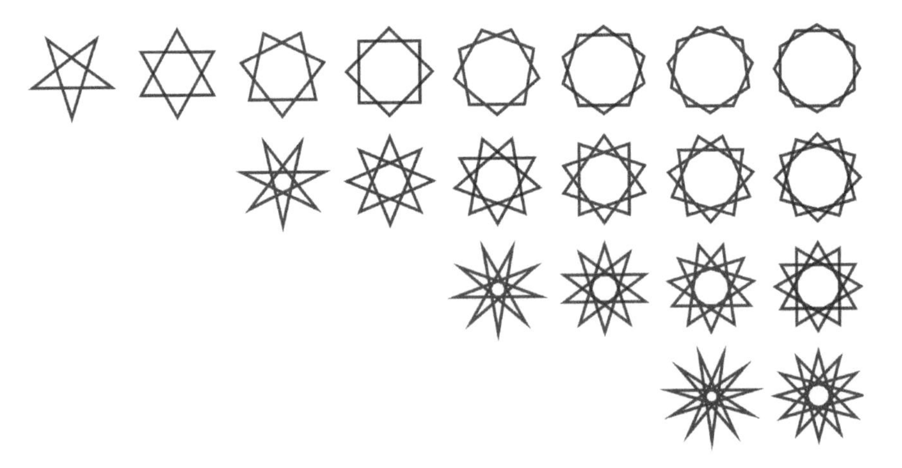

Figure 2.4: Stellating n-gons from n = 5 to 12. Figure 2.8, p. 11, shows more stellations. Some of these are considered to be star polygons.

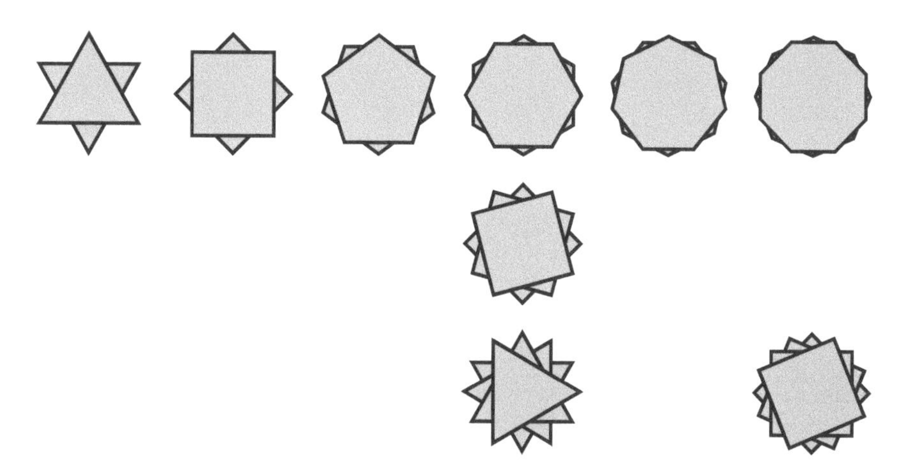

Figure 2.5: Top row: pairs of overlapping regular n-gons from n = 3 to 8. Second and third rows: more than two n-gons can be overlapped, e.g. three squares, four equilateral triangles and four squares.

2.4 Joining the diagonals of a regular polygon to make a star polygon

Take a regular octagon. For each vertex, make a line joining it with the vertex one away from the adjacent vertex. In other words, connect each vertex with every second vertex. This makes an octagram, but joining every third vertex makes a different octagram (figure 2.6). The former is considered to be degenerate as more than a single thread is needed to wind around the vertices.

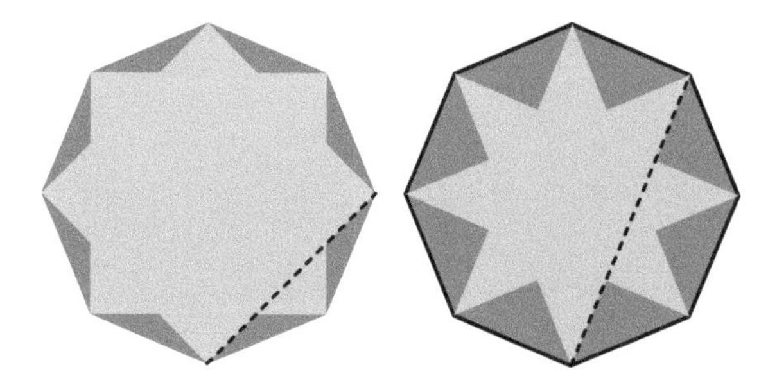

Figure 2.6: Stars made by joining every third vertex (left) and fourth vertex (right) of a regular octagon.

These octagrams are *star poylgons*. Although they look different to the more familiar convex regular polygons (the edges intersect), star polygons are considered by some to be regular polygons as all edges have the same length and all vertices have the same interior angle. Only the outermost vertices are considered: the crossings nearer the centre are not considered to be vertices.

Star polygons can also be created with turtle graphics. For example, in Logo:

```
TO POLYGON :length :sides :repeat
REPEAT :repeat [
```

```
FD :length
RT 360/:sides
]
END

HOME
CS
POLYGON 100 5/2 7
```

In fact, the Schläfli symbol for this example is $\{\frac{5}{2}\}$ (figure 2.7). The numerator 5 specifies the number of points in the star polygon. The denominator 2 defines the density of the star polygon: a ray from the centre going outward will pass through two lines of the star polygon. Table 2.2 shows the interior angle of selected star polygons and some of these are shown in figure 2.8. Both omit cases that look similar, e.g. $\{\frac{5}{2}\}$ looks the same as $\{\frac{5}{3}\}$ or have zero area, e.g. $\{\frac{6}{3}\}$.

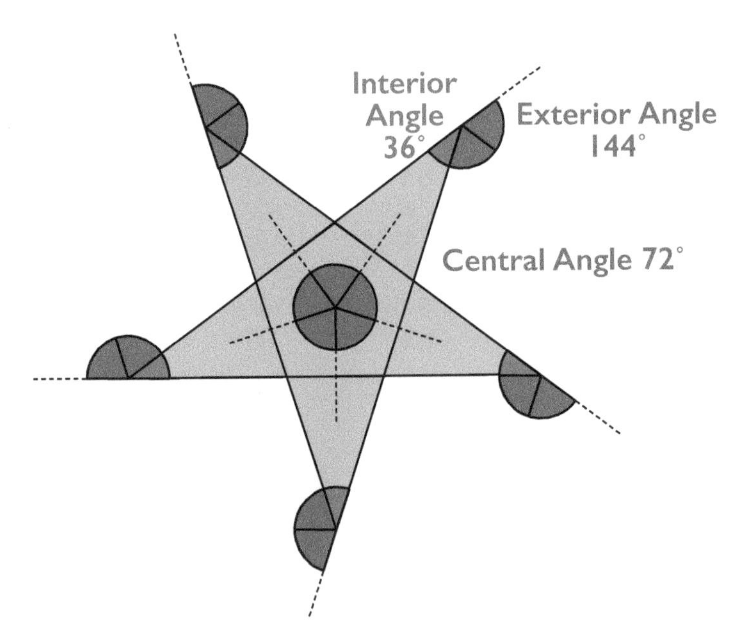

Figure 2.7: Star polygon with Schläfli symbol $\{\frac{5}{2}\}$. The central angle is 360° divided by the number of sides. The exterior angle is $\frac{2}{5}$ of 360°. The sum of the exterior and interior angles is 180°.

For a non-degenerate star polygon the fraction in the Schläfli symbol needs to be in its lowest terms (a reduced fraction): the numerator and denominator are coprime, i.e. they have a greatest common divisor of 1, e.g. $\{\frac{3}{8}\}$. We have seen that $\{\frac{2}{8}\}$ is degenerate.

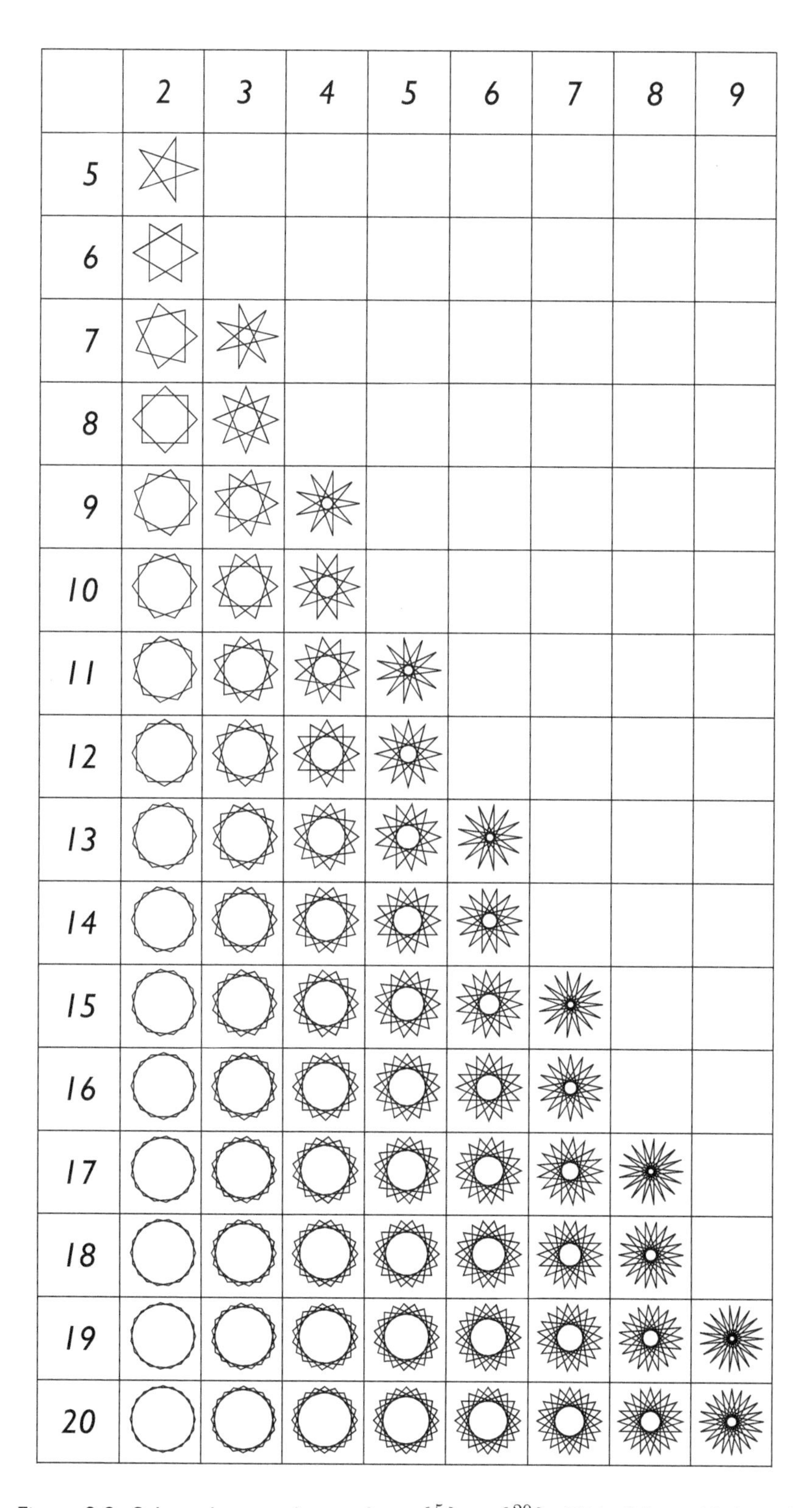

Figure 2.8: Selected star polygons from $\{\frac{5}{2}\}$ to $\{\frac{20}{9}\}$. Table 2.2, p. 12, lists the interior angle of the outermost arm of the star.

Table 2.2: Interior angles of selected regular star polygons $\{\frac{e}{n}\}$ given up to 3 decimal places. Figure 2.8, p. 11 shows the star polygons.

n	e							
	2	3	4	5	6	7	8	9
5	36							
6	60							
7	77.143	25.714						
8	90	45						
9	100	60	20					
10	108	72	36					
11	114.545	81.818	49.091	16.364				
12	120	90	60	30				
13	124.615	96.923	69.231	41.538	13.846			
14	128.571	102.857	77.143	51.429	25.714			
15	132	108	84	60	36	12		
16	135	112.5	90	67.5	45	22.5		
17	137.647	116.471	95.294	74.118	52.941	31.765	10.588	
18	140	120	100	80	60	40	20	
19	142.105	123.158	104.211	85.263	66.316	47.368	28.421	9.474
20	144	126	108	90	72	54	36	18

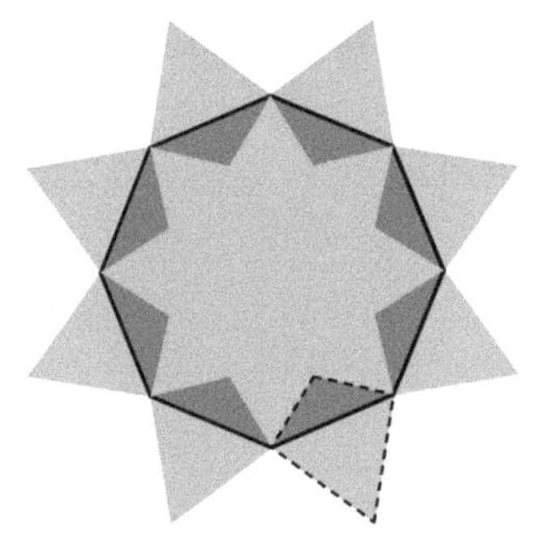

Figure 2.9: Moving the midpoints of the edges of a regular octagon makes the two different stars in this example.

2.5 Taking the midpoints of the edges of a regular polygon and moving them towards or away from the centre

Figure 2.9 shows a couple of possible examples for a square. For a square, if the midpoint is taken towards the centre and then beyond to meet the midpoint of the opposite edge, this eight-pointed star is sometimes called the Brunés star. Tons Brunés was a Danish engineer who believed that a system of ancient geometry was the basis of many ancient temples [Nicholson et al. 98].

2.6 Rosette

This method builds on the previous method. Figure 2.10 shows four stages of one rosette.

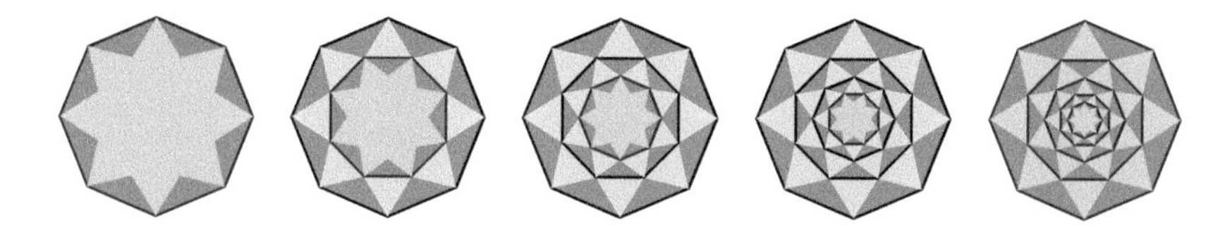

Figure 2.10: Making a rosette from an eight-pointed star.

1. Find the midpoints of the edges of a regular polygon.
2. Move these points towards the centre.
3. Create a new polygon using these points.
4. This creates a polygonal ring of triangles: we can make a copy and scale and rotate it to fit inside. We are interested in the scale factor of the inside edge compared with the outside edge. This scale factor should be convenient, e.g. $\frac{1}{2}$, $\frac{1}{\sqrt{2}}$, $\frac{1}{\sqrt{3}}$, etc. Also, the angles should be convenient, e.g. 45°, 60°, 72°, etc. If not, we can combine two rings and have kites instead of triangles.

Experimenting with this method produces some classic cases (figure 2.11):

- Diminishing kites in an octagon.
- Diminishing squares in an octagon.
- Diminishing kites composed of isosceles right-angled triangles (IRATs) and equilateral triangles in a dodecagon.

Figure 2.11: Left: diminishing kites in an octagon. Centre: diminishing squares in an octagon. Right: diminishing isosceles right-angled triangles (IRATs) and equilateral triangles in a dodecagon.

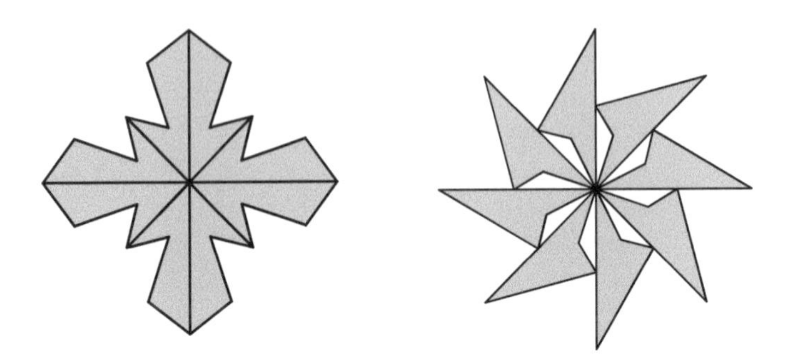

Figure 2.12: Left: two mirrors placed at 45°make a simple kaleidoscope. Right: an eight-pointed pinwheel is made by rotating a shape about a centre of rotation.

2.7 Kaleidoscope mirrors

This is a flexible method that makes many star shapes. Take two mirrors and make a V-shape. If the angle is 45°, the mirrors can make star patterns like figure 2.12 (left). If you add a third mirror to make an isosceles right-angled triangle then you have made one kind of kaleidoscope. Another type of kaleidoscope has mirrors arranged in an equilateral triangle.

2.8 Taking a shape and rotating it around a centre of rotation

This is another flexible method that makes many star shapes. Take any shape and choose a whole number that is at least three. Choose a centre of rotation and make n copies of the shape, rotating each by a multiple of $\frac{360^\circ}{n}$. The centre of rotation does not have to be located on the original shape. Figure 2.12 (right) shows an eight-pointed pinwheel.

Chapter 3

Special Rectangles

This book uses rectangular paper which includes squares: a square is a rectangle whose sides are all equal. Irregular rectangles are sometimes called *oblongs* to distinguish them from squares.

Figure 3.3 summarises some special rectangles. It builds on a summary by Dave Mitchell [Mitchell 15]. Using these rectangles means that the angles and lengths needed to make specific stars are easy to obtain. Also, there is little waste from hiding many layers of paper away.

Figure 3.3 shows:

Name This is the common name of the rectangle with the given proportions. Sometimes it is the name used by paperfolders.

Angle between diagonal and long side The diagonal of a rectangle is the line segment connecting opposite corners (vertices). The angle between the diagonal and the long side of the rectangle is important as it is easy to make.

Self similarity A shape is *congruent* with another shape if they are identical: they coincide exactly when superimposed (figure 3.1). Shapes are geometrically *similar* if they have the same angles and proportions but are of different sizes (figure 3.2). *Self-similarity* means that a shape is similar to a copy of itself on a different scale. For example, a 2:1 rectangle contains four similar 2:1 rectangles: the smaller 2:1 rectangles have dimensions 0.5:1 and can either be arranged as a 2 by 2 grid, or more interestingly, in a single row.

Common examples The silver rectangle is familiar because the international standard A4 paper has the same proportions. Some other examples are given.

Some other properties Some rectangles fit inside regular polygons in special ways. For example, you can overlap three bronze rectangles to make a regular hexagon. You can see a six-pointed star if you use translucent paper or view the hexagon in front of a strong light.

DOI: 10.1201/9781003184492-3

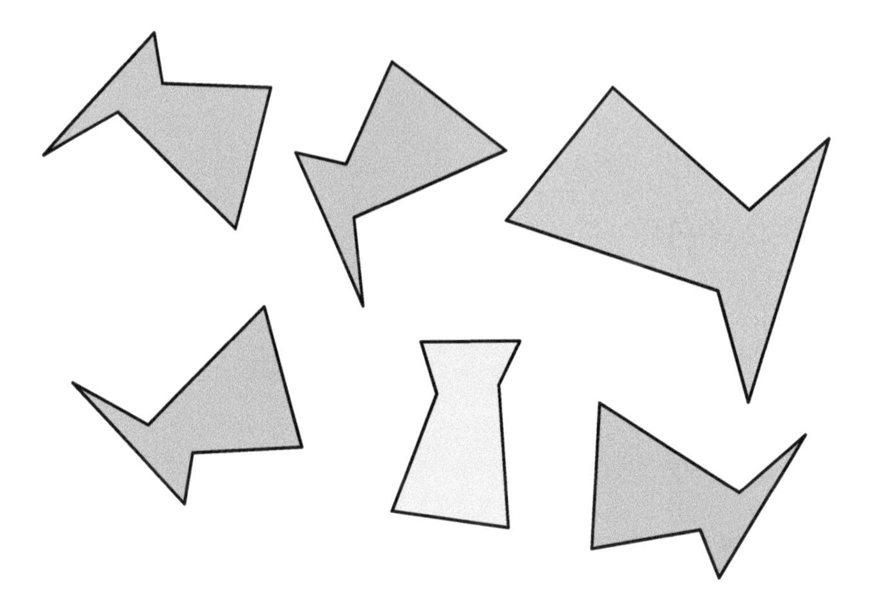

Figure 3.1: The blue shapes are *congruent*: they are are identical and they coincide exactly when superimposed. In two dimensions, mirror images of a shape are considered to be congruent.

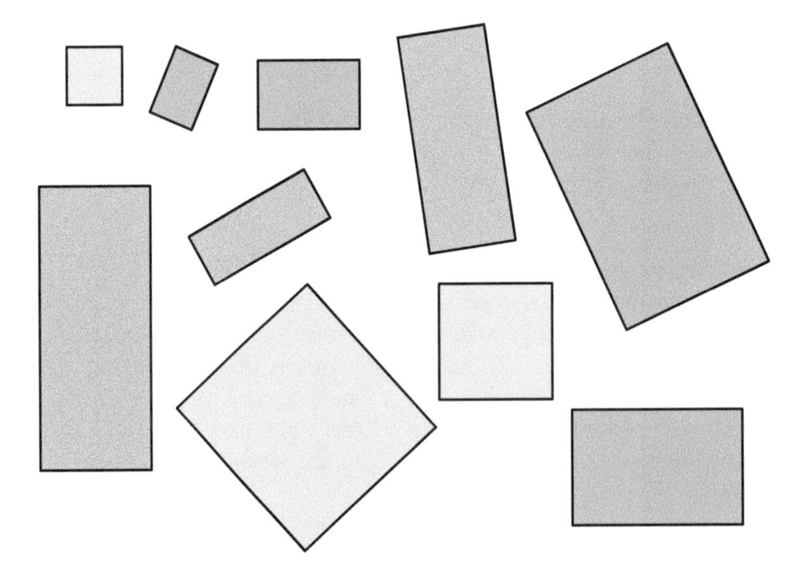

Figure 3.2: Shapes are geometrically *similar* if they have the same angles and proportions but are of different sizes. Here the blue rectangles are similar, the red rectangles are similar and the squares are similar.

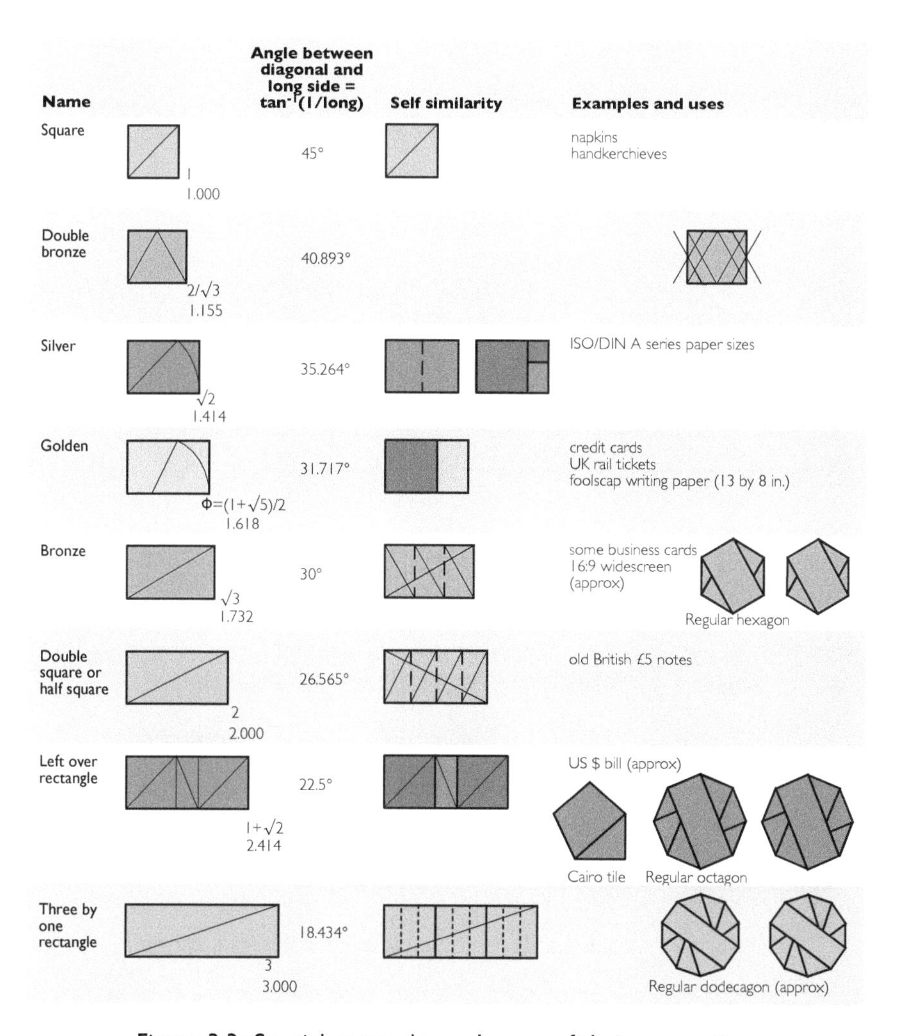

Figure 3.3: Special rectangles and some of their properties.

3.1 The square

The square is a familiar regular polygon that has four equal sides and four vertices of 90°.

Making a square from an oblong requires you to bisect a corner by folding the shorter edge onto the longer edge (figure 3.4). You have now found the point on the longer edge that makes a square.

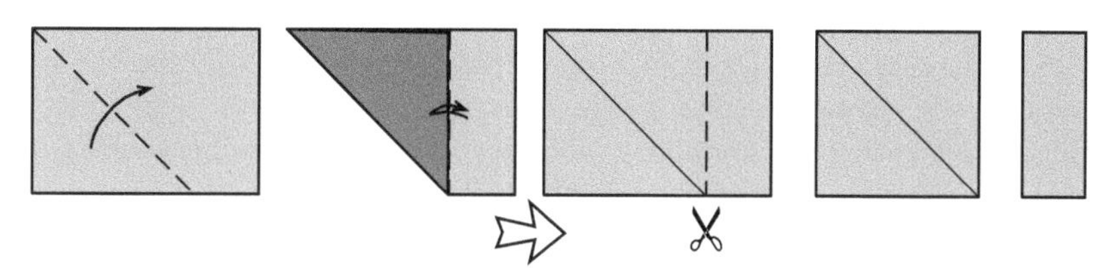

Figure 3.4: Making a square from an oblong.

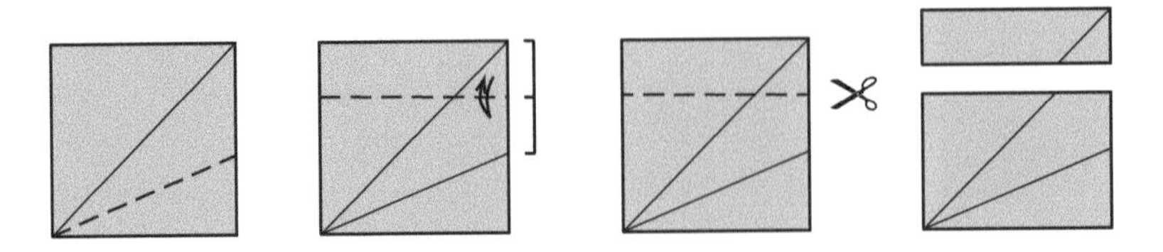

Figure 3.5: Making a silver rectangle from a square.

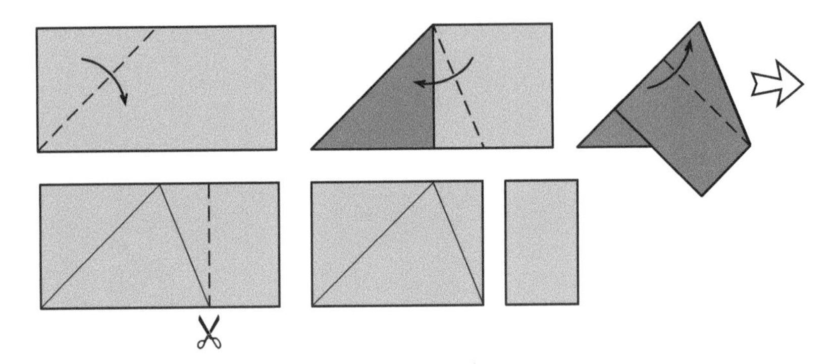

Figure 3.6: Making a silver rectangle from a long rectangle.

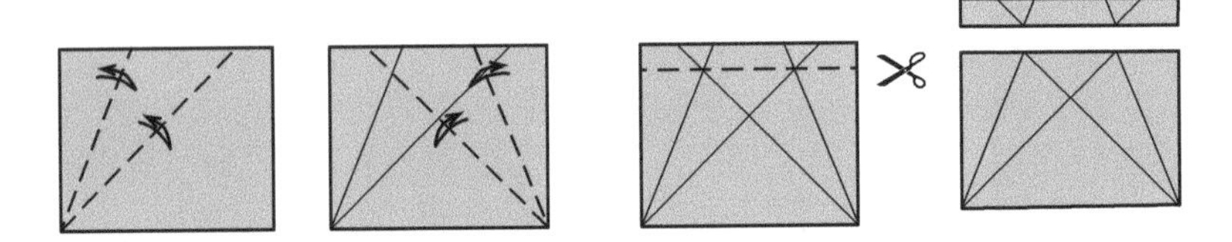

Figure 3.7: Making a silver rectangle from a short rectangle.

A unit square has sides that are one unit long; the diagonal has length $\sqrt{2}$. Cutting a square along the diagonal makes two isosceles right-angled triangles (IRATs). Each IRAT can be cut in half to make two similar IRATs: figure 12.12 (p. 120) shows this and triangles that dissect into three and five smaller similar triangles. Any triangle can be dissected into four smaller similar triangles using the midpoints of the edges.

3.2 The silver rectangle

A rectangle whose sides are in the ratio $1 : \sqrt{2}$ is known as a *silver rectangle* to origami enthusiasts. Like the IRAT, a silver rectangle can be cut into two shapes that are similar to the original shape.

Making a silver rectangle from a square requires you to bisect the 45° angle of an isosceles right-angled triangle by folding the shorter lower edge onto the hypotenuse (figure 3.5). You have now found the point on the hypotenuse that makes the lower edge the longer side of a silver rectangle.

Making silver rectangles from non-silver rectangles is fairly straightforward. The diagonal of a unit square is $\sqrt{2}$, so it's easy to find the point that defines a silver rectangle. In practice there are three methods: one for a rectangle longer than a silver rectangle (figure 3.6), one for a rectangle shorter than a silver rectangle (figure 3.7) and one for a square (figure 3.4).

Paper from the international standard A series are silver rectangles (figure 3.8). This standard defines A0 to have an area of 1 m² and sides in the ratio $1 : \sqrt{2}$. A1 has half the area of A0 but also has sides in the ratio $1 : \sqrt{2}$. A2 has half the area of A1, and so on. The common A4 paper has sides 210 by 297 mm (8.27 in by 11.7 in). US letter paper is 8.5 in by 11 in (215.9 mm by 279.4 mm) so is about 6 mm (0.24 in) wider and 18 mm (0.71 in) shorter than A4. To convert US letter to a silver rectangle, trim a long strip $\frac{3}{8}$ in (9 mm) wide.

The diagonal of a silver rectangle makes an angle of about 35.2° with the long side. This is close to 36° which means that it is easy to approximate regular pentagons. The error is small and negligible for small sheets of paper. However, for larger sheets where the error would be noticeable, the rectangle should be $1 : tan(36°)$ which is $1 : 1.376$ to 3 decimal places. This means 2.675% should should be trimmed to make the sheet shorter. For an A4 sheet, this is an 8 mm strip.

For some ideas on demonstrating and proving the proportions of silver rectangles, see Explore the properties of A series paper (p. 117).

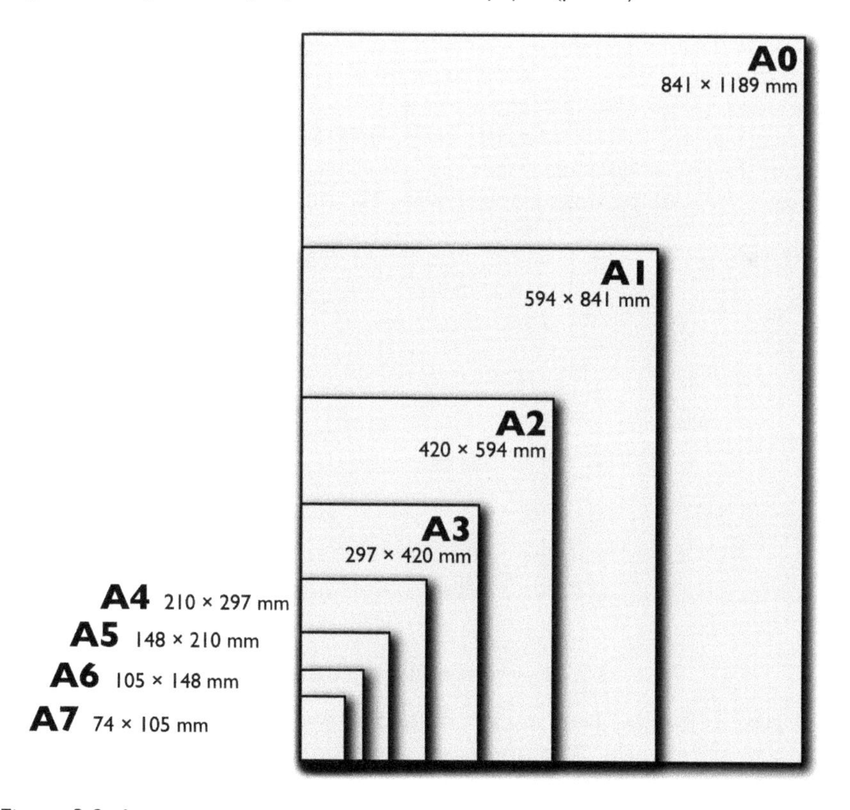

Figure 3.8: International A series paper sizes. A0 is defined to have an area of 1 m^2 and sides in the ratio $1 : \sqrt{2}$.

3.3 The leftover rectangle

If you cut a square off a silver rectangle, the result is the *leftover rectangle* (LOR). It has sides in the ratio $1 : 1 + \sqrt{2}$. If you cut *two* squares off an LOR, the remaining rectangle is similar to the original LOR.

The diagonal of the LOR makes an angle of 22.5° with the long side. You can form a regular octagon from four congruent LORs by laying them down so that the short edges of the LOR form the opposite edges of the regular octagon.

Confusingly, some mathematicians call the LOR a silver rectangle because they call the ratio of sides, $1 : 1 + \sqrt{2}$, the silver mean.

3.4 The bronze rectangle

A rectangle with a diagonal 30° to the long edge is known as a bronze rectangle. Its sides are in the ratio $1 : \sqrt{3}$. The bronze name continues the metallic theme of the golden and silver rectangles.

The bronze rectangle can be cut into three similar bronze rectangles. You can form a regular hexagon from three congruent bronze rectangles by laying them down so that the short edges of the bronze rectangles form the opposite edges of the regular hexagon.

The usual method for making a bronze rectangle uses a fold that trisects a right angle (figure 3.9). First fold the long edges together and unfold. Then make a fold that goes through the lower left corner, and at the same time, place the right lower corner onto the centre crease. Start with a soft fold first, then keep adjusting until the fold is in the right place. Finish with a firm crease. A fold that satisfies the two conditions divides the lower left right angle into 60° and 30°. For some ideas on proving this, see Folding 60° angles (p. 124).

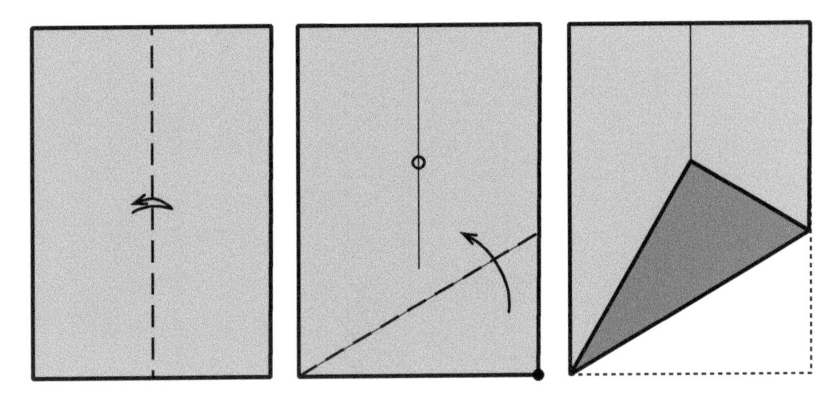

Figure 3.9: How to trisect a right angle by folding.

Figure 3.10 shows how you can make one, two or three bronze rectangles from a silver rectangle. A square makes two bronze rectangles with little waste (figure 3.11).

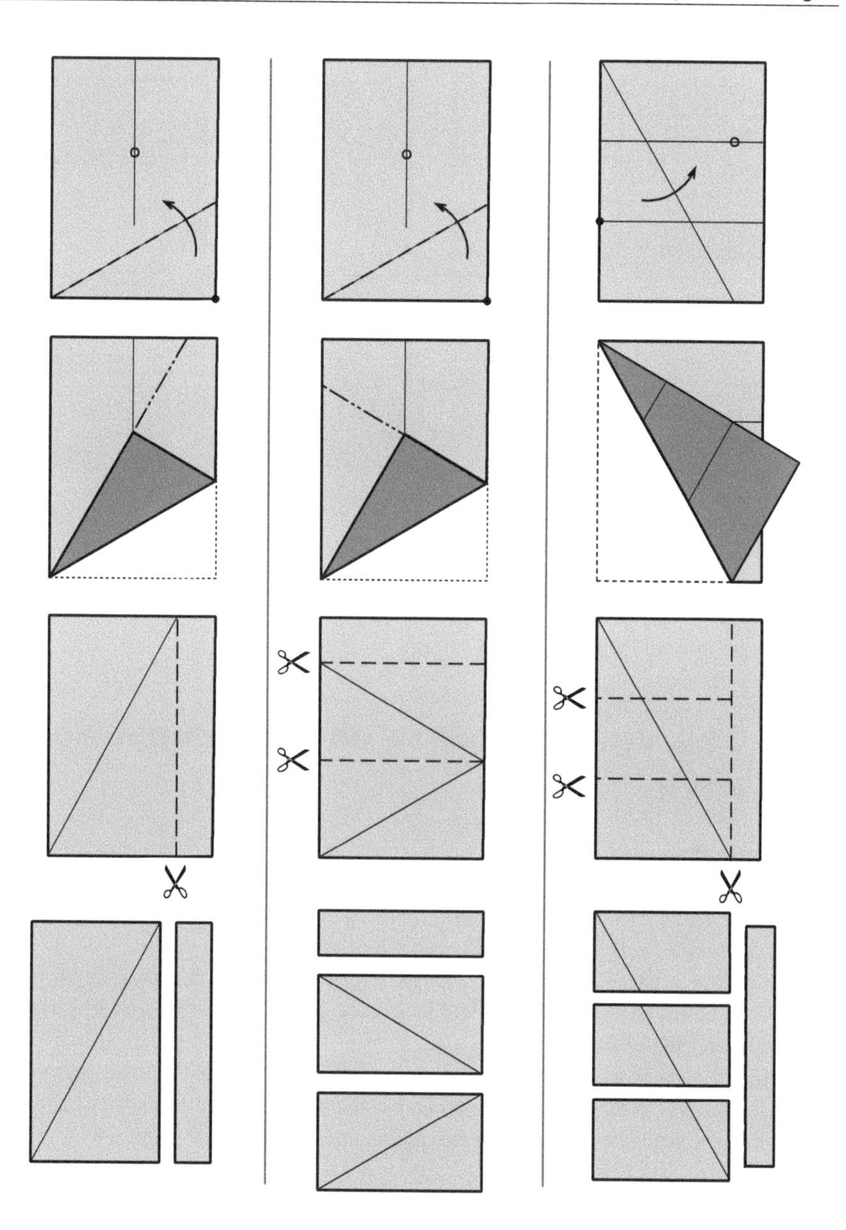

Figure 3.10: How to make one, two or three bronze rectangles from a silver rectangle.

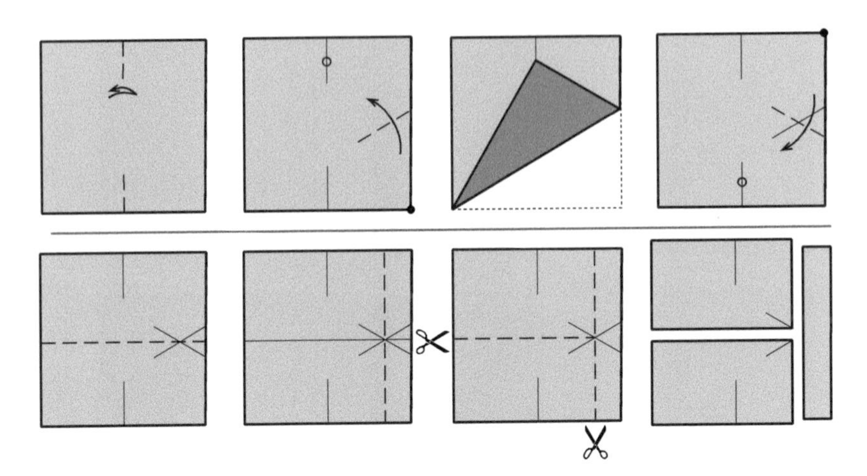

Figure 3.11: How to make two bronze rectangles from a square. For a double bronze rectangle, only cut away the small tall strip on the right, or fold it out of the way. Partial 60° folds are shown but you may want to make full creases if they are needed in your chosen project.

3.5 The double bronze or half bronze rectangle

The double bronze rectangle has sides in the ratio of $2 : \sqrt{3}$. It can be made from a square with little waste by adapting the sequence in figure 3.11. This means that some models can be made from squares without cutting, e.g. *Star 6, Bronze Rectangle from Square* (p. 77).

When the greatest equilateral triangle is put inside the double bronze rectangle, the apex is the midpoint of one long edge of the rectangle, and the base is the opposite long edge.

When the greatest regular hexagon is put inside the double bronze rectangle, two opposite vertices of the hexagon lie on the midpoints of the short edges of the rectangle, and two sides of the hexagons lie on the longer edges.

3.6 The golden rectangle

The diagonal of a $1 : 2$ rectangle has a length of $\sqrt{5}$. This leads a way to the golden ratio, $\frac{\sqrt{5}+1}{2}$, 1.618 (and also 0.618, $\frac{\sqrt{5}-1}{2}$) (figure 3.12).

The golden rectangle is famous in mathematics but is rarely used in origami. It has the property that if a square is cut off, the remaining oblong is also a golden rectangle.

You can fold some of the models in this book from golden rectangles, e.g. *Starfish* (p. 105) and *Star 6, Equilateral Triangle from Rectangle* (p. 87). However, they do not exploit the geometry of the golden rectangle.

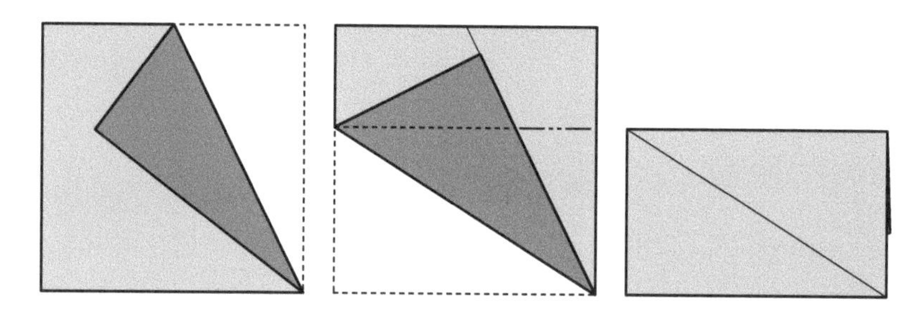

Figure 3.12: Folding a golden rectangle from a square uses the diagonal of the 2:1 rectangle embedded inside the square.

3.7 The double square or half square

The 2:1 rectangle can approximate angles in regular polygons with 5, 7 and 10 sides.

Figure 3.13 shows two of the ways a 2:1 rectangle can approximate angles of a regular pentagon:

- The diagonal makes an angle of 26.565° with the long edge. Twice this angle is 53.130° which is close to 54°, i.e. three fifths of 90°. This is half of the interior angle of a regular pentagon.
- The diagonal makes an angle of 63.435° with the short edge. Add 45° to make 108.435°.

The double square rectangle has sides in the ratio of $1 : 2$. It's easy to cut a square in half to make two: they are half-squares. A 1:2 rectangle can be cut into a row of four smaller 1:2 rectangles.

Folding a square in half by joining opposite edges together is known as a book fold. It makes two 2:1 rectangles hinged on the long edge that makes approximating decagonal angles convenient. Figure 3.14 shows one unit inserted into another: 143.130° approximates the interior angle of a regular decagon, 144°. This is used by Nakano's *Flying Saucer 4* [Nakano 85, p. 61] and Max Efremov's *Ornament 2* [Afonkin and Hull 98, p. 16], although the latter blintzes the square (folding all the corners to the centre) first to create extra pockets. *Decagon Ring* (p. 65) adapts Nakano's model to make a ring that holds together without using glue.

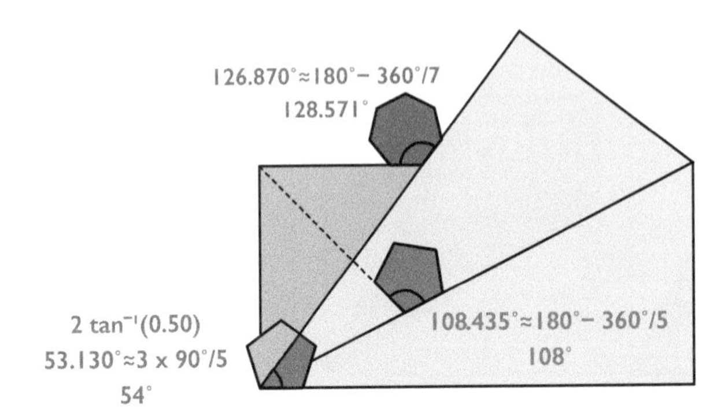

Figure 3.13: Folding the diagonal of a 2:1 rectangle gives approximate angles needed for regular heptagons and pentagons.

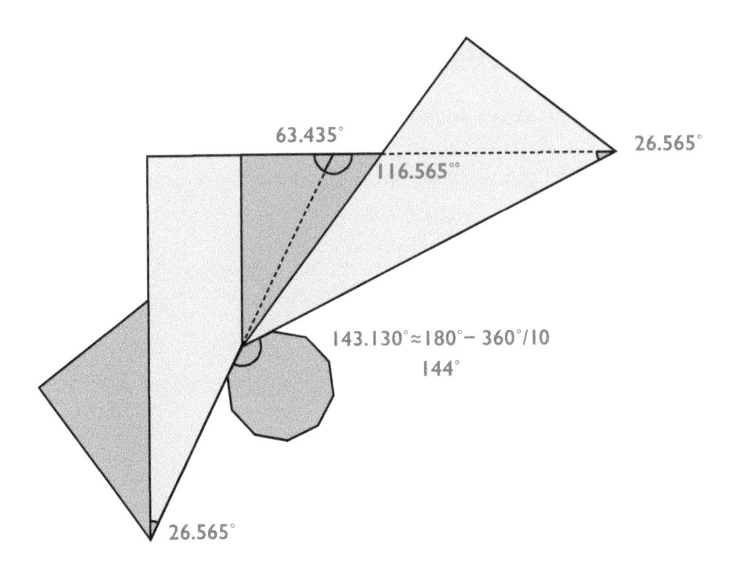

Figure 3.14: Fold a square in half then fold on the diagonal of the 1:2 rectangle. Joining two units approximates the interior angle of a regular decagon.

The 2:1 rectangle can also approximate the interior angle of a regular heptagon, 128.571°. When the rectangle is folded along the diagonal, the two long edges make an angle of 126.870° (figure 3.13). Both Beynon and Meeusen used this for seven-piece models (*Heptad Ring* [Beynon 90, p. 12] and *Seven Module Crown* [Palacios 00, p. 114], respectively). Petty embedded the 2:1 rectangle inside a square as a 0.5:0.25 rectangle for his *Septadring* [Petty 98, p. 27]).

3.8 The 1:3 rectangle

A $1:3$ rectangle has a diagonal that approximates 36°. Cutting three 1:3 rectangles from a square requires dividing an edge into thirds: figure 4.9 (p. 34) shows several methods. Instead of folding a 1:3 rectangle, you can embed a 1:3 rectangle inside a larger rectangle that has more convenient dimensions. As with the silver rectangle approximation of 36°, the error is small and negligible for small sheets of paper. However, for larger sheets where the error would be noticeable, the rectangle should be $1 : tan(18°)$, $1 : 3.078$. This means the 1:3 rectangle should be 2.589% taller.

To divide the bottom right right angle of a unit square in the ratio two to three, you could fold the bottom edge to point (0.75, 0.75) whilst creasing through (1, 0). However, figure 4.11 (p. 36) shows a more accurate method that needs only a few more creases. This locates the start of the crease on the x-axis with only 0.277% error. It uses the fact that $tan(36°) \approx \frac{\frac{3}{4}+\sqrt{\frac{1}{2}}}{2}$, i.e. $0.7265 \approx 0.7285$. This means we need to crease halfway between $\sqrt{\frac{1}{2}}$ and $\frac{3}{4}$, i.e. 0.7071 and 0.7500. 0.7071 is the midpoint of the corner and the meet of the kite fold.

Another way to use a 1:3 rectangle to approximate pentagons is to divide it into a square and a 1:2 rectangle. Figure 3.15 shows that adding 45° to the angle between the short side and the diagonal of the 1:2 rectangle approximates 108°. *Star 5, 3 by 2 from Square* (p. 51) and *Star 5, 3 by 2 from Square Blintz* (p. 52) use this approximation. Robert Neale embedded a 1:2 rectangle inside a 1:4 rectangle to make a unit for a dodecahedron (the 1:4 rectangle is a square pleated into quarters) [Plank 19].

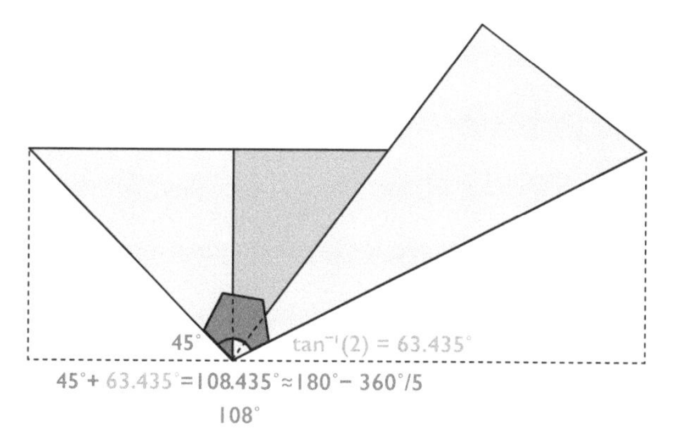

Figure 3.15: A 1:3 rectangle contains a square and a 1:2 rectangle. The angle between the diagonals approximates the interior angle of a regular pentagon.

3.9 The 4:5 rectangle

The diagonal of a 4:5 rectangle makes an angle with the long edge that approximates 360°÷7.

3.10 Landscape-portrait (LP) rectangles

The oblongs used for *Poly Diag Star 5* (p. 54), *Poly Diag Star 6* (p. 53) and *Poly Diag Star 8* (p. 55) have some interesting properties. They consist of a rectangle in portrait orientation (standing on the short edge of rectangle) and similar rectangle in landscape orientation (standing on the long edge of the rectangle), hence the name LP rectangle (figure 3.16).

The ratio of the sides of the whole LP rectangle is $n : (1 + n)$ or $1 : \frac{1}{n} + 1$. The long edge of the whole LP rectangle can be the hypotenuse of a right-angled triangle: the two other sides of the triangle are the diagonals of the component landscape and portrait rectangles.

When n is of the form $\sqrt{i}$, we can cut $i+1$ LP rectangles from a rectangle whose sides are in the proportion $1 : \sqrt{i}$. For example, the right side of figure 3.16 shows that a silver rectangle can be cut into three LP rectangles.

Poly Diag Star 8 (p. 55) uses two LP rectangles made from a silver rectangle cut along the long mirror line. n is $1 : 1+\sqrt{2}$.

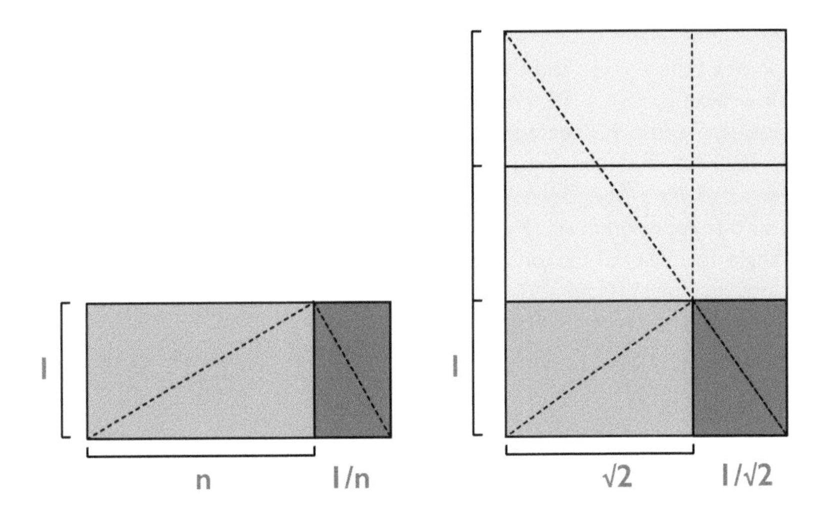

Figure 3.16: Left: the blue rectangles form a landscape-portrait rectangle. Right: the proportion of each blue rectangle is $1 : \sqrt{2}$. The blue rectangles form an LP rectangle of proportion $\sqrt{2} : (1+\sqrt{2})$ or $1 : \frac{1}{\sqrt{2}} + 1$. Three such LP rectangles can be cut from a silver $(1 : \sqrt{2})$ rectangle.

3.11 Other rectangles

Kawamura gave more examples like 1:3 and 4:5 for regular n-gons with n between three and ten [Kawamura 03]. For example, the late Francis Ow embedded an 11:8 rectangle in a square for his *Pentagon Coaster (Dodecahedron Unit)* [Biddle and Biddle 93, p. 21]. Beynon made a regular 13-gon using a 1:2 rectangle folded in half (i.e. a 1:4 rectangle) for his *Lucky Ring* [Beynon 91, p. 53].

Chapter 4

Folding and Generalising Stars

As you fold the models in this book, you'll notice some similarities between some models. These models are part of a series or are related to each other. Other models are not directly related, but you may wonder if they could be varied or generalised.

This chapter explains some ways to vary and generalise modular origami stars. I have tried to strike a balance between showing how a model can be generalised and giving you scope to explore these possibilities yourself.

4.1 Variation and creation

In general, I have only included successful variations as some do not work well, e.g. the units of variation do not join well as the original. However, you may well find effective variations yourself.

You may find that you have created something that has already been created by someone else. This experience is not a rare experience, especially with simple geometric origami. For example, *Pinwheel Square Slider* (p. 95) is similar, but not identical to a model by Francis Ow [Ow 18a]. I only found Ow's model when this book was almost finished. *Intersecting Squares* (p. 89) is almost the same as model "22 by unknown" [Jackson 82].

Please do not be put off as you may come up with something new that makes other people say "Why didn't I think of that?" Another way of accepting the situation is to acknowledge that you have trodden the same creative path as others.

It is a courtesy to give credit to other people if you derive variations of their work (just as you'd like other people to acknowledge your work if they use it). If your creation is sufficiently different, then you can claim your work as your own. How different does yours have to be? It's a matter of personal ethics: usually the difference has to be non-trivial or change the result or process in an unexpected way. Even so, a credit to the original creator is respectful and can help others in tracing how ideas spread and develop.

4.2 Varying and generalising a flat star

You usually have a few methods for varying and generalising a flat star. Not all methods are possible for a given star.

DOI: 10.1201/9781003184492-4

Figure 4.1: *Spinning Star 5* made from five units of *Intersecting Squares* (p. 89).

4.2.1 Changing the number of units without changing the unit

This is rare for 2D models and is more likely for 3D models. The number of units for *Starfish* (p. 105) can be varied without changing the key angle. However, there is a minimum number of units.

Joining between three and seven units for *Intersecting Squares* (p. 89) will make 3D stars. Figure 4.1 shows five units. The units need to be symmetrical. To make the model spin, cup your hands and gently hold the model in your palms. Blow to spin.

4.2.2 Embedding in a different rectangle without changing the key angle

Sometimes it's possible to change the proportions of the paper but keep the existing geometry. This changes the lengths and how the units attach, but not key angles and number of units. For some 3D stars, there are other choices.

For some models, the simplest variation is to lengthen the rectangle. This is usually possible for models with parallel sides, e.g. *Square Pointer Slider* (p. 94). Extra paper is added to the middle: the left and right ends of the unit stay the same.

Doing this to a 2D model can make a 3D version, e.g. *Pinwheel Square Slider* (p. 95) becomes *Pinwheel Square 3D Slider* (p. 96). This technique is known as a strip graft [Lang 11, p. 141]. Other kinds of graft exist: e.g. the graft can be one-sided or go all around the original rectangle (border graft [Lang 11, p. 135]).

4.2.3 Embedding in a different rectangle and changing the key angle

Change the proportions of the paper and, in turn, change the lengths and angles of the units. The key angle of *Starfish* (p. 105) is between the diagonal and the edge of the rectangle. Figure 4.2 shows the effect of using different proportions for the starting rectangle. As the rectangle becomes longer, the small locking flaps at the top right and bottom left become smaller. They disappear when the rectangle is bronze.

Rectangles longer than bronze rectangles have extra flaps to be folded away: these are at the top and bottom of the rectangle. When the rectangle is a leftover rectangle, these flaps align with the folded edges. The next rectangle where the flaps align is the 1:3 rectangle. In fact the key rectangles of bronze, leftover and 1:3 divide the top left and bottom right corners into three, four and five equal angles, respectively.

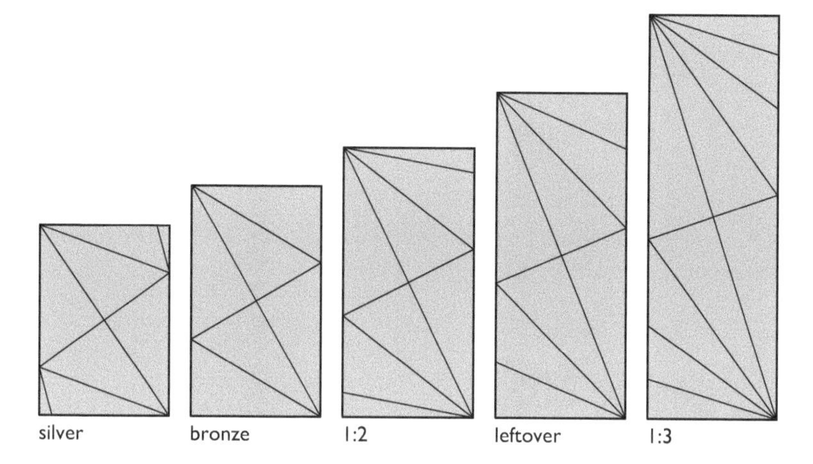

Figure 4.2: The crease patterns of the Starfish unit from rectangles of different proportions.

4.3 Varying and generalising a single model

4.3.1 Varying Star 8 Octad

We now explore the two previous techniques for varying *Star 8 Octad* (p. 68).The first step is to book fold a rectangle, i.e. fold in half bringing opposite edges together. The next step determines the angle and the number of units. For *Star 8 Octad* the fold starts from the lower left corner and reaches 0.707 up the right edge. The right location point for *Nonad Star 9* is 0.75 up the right edge. The diagonal is inside a 3:8 rectangle instead of a leftover rectangle (figure 4.3).

The locking flap is small so the assembly is weak. *Nonad Star 9 Ribbon* (figure 4.4) has a larger locking flap by valley folding the edge of the square. In some ways the unit is easier to make but the arrangement of inner triangles might be considered to be less pleasing.

Instead of folding the edge of the rectangle for a larger locking flap, we can eliminate it so that the key crease is a diagonal (figure 4.5, right). Five silver rectangles make *Pentad Star 5* and six double bronze rectangles make *Hexad Star 6*.

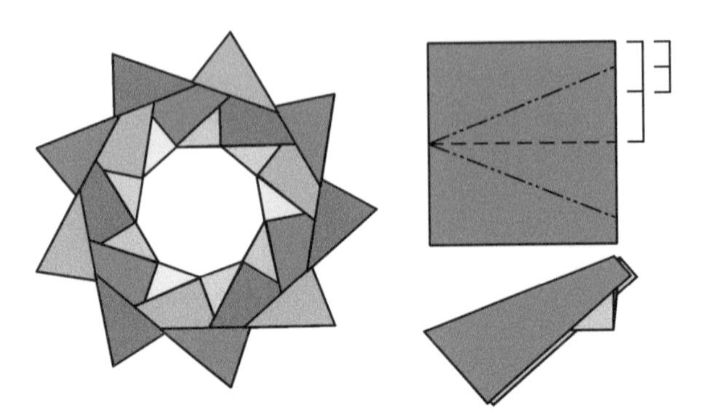

Figure 4.3: *Nonad Star 9* can be made from nine squares. After folding in half, the location point is three quarters up the right edge.

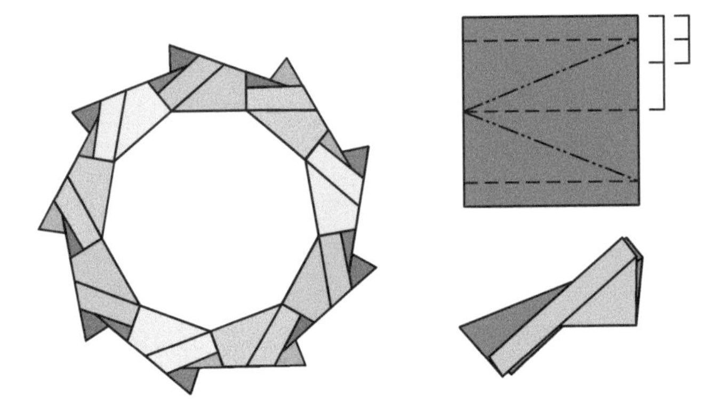

Figure 4.4: *Nonad Star 9 Ribbon* adds two extra folds to unit in figure 4.3. The larger locking flap makes for a stronger lock.

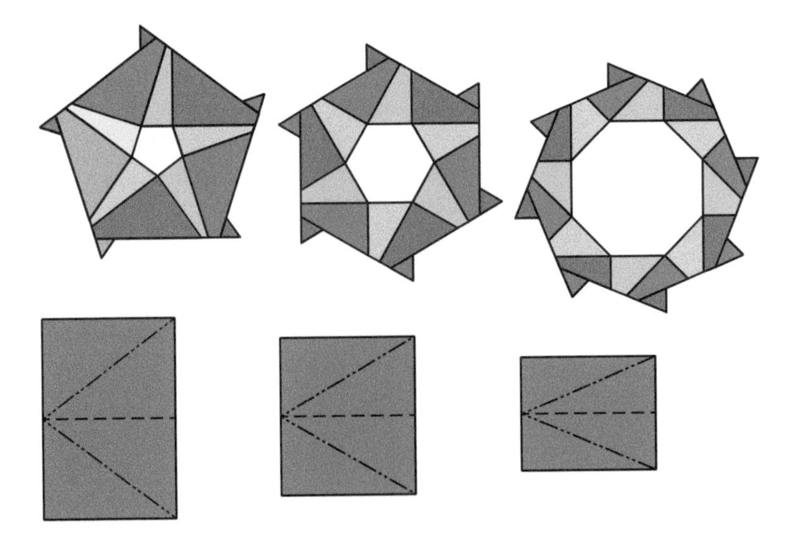

Figure 4.5: *Star 8 Octad* (p. 68) can be made from a double leftover rectangle instead of a square: the module's main creases are shown below (right). Left and middle show the unit made from a silver rectangle and a double bronze rectangle, which make a *Pentad Star 5* and *Hexad Star 6*, respectively.

Starburst 6 (p. 71) embeds the double bronze rectangle of *Hexad Star 6* (p. 30) inside a square. It grafts extra paper on one side.

It exploits another possibility: the units hold together even though the front and back are not folded in the same way. A further variation leads to the eight-piece *Octad Starburst 8 60°* (p. 70).

4.3.2 Six types of variation from a single model: Star 4, Square From Silver Rectangle

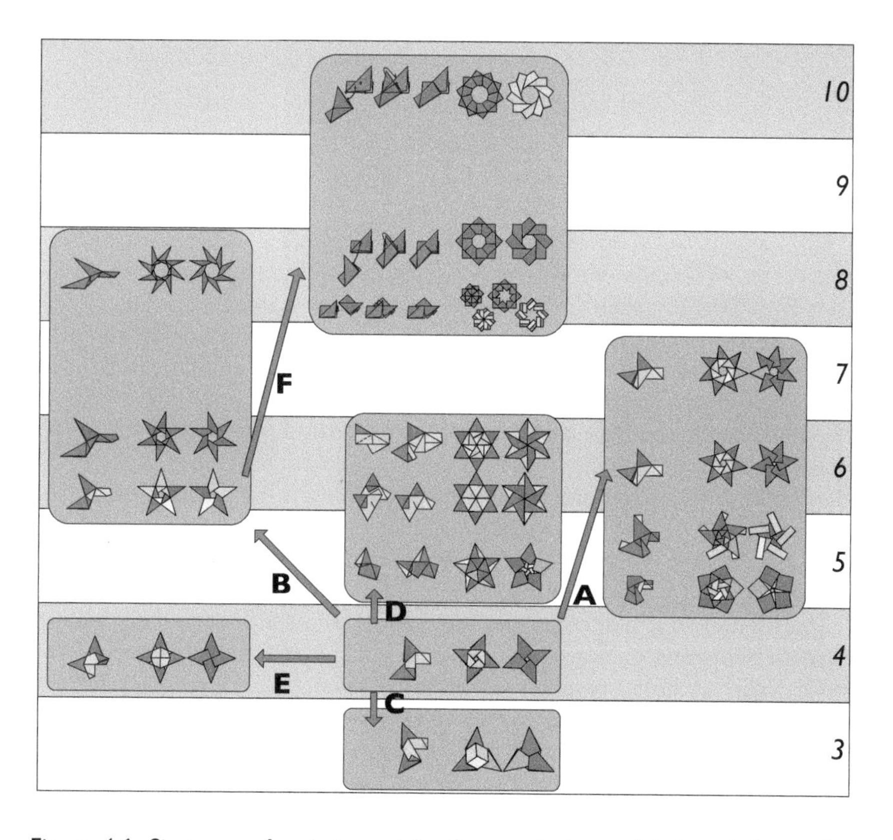

Figure 4.6: Six types of variation arising from a single model, *Star 4, Square From Silver Rectangle* (p. 46). A: change the angle at the point of attachment. B: use landscape-portrait rectangles to change the angle at the tip of the star's arms. C: allow a flap to project beyond the arm of the star. D: change the first fold from 45° to 60° or another angle. E: change the second fold to be at an angle to the edge of the square so that the arm of the star is symmetrical. F: tuck the flap between the layers of the square instead of placing on top.

Figure 4.6 shows some of the possibilities arising from a single model. Some of these connections were made at the time of creating; others were made in retrospect. *Star 4, Square From Silver Rectangle* (p. 46) is a simple and effective four-piece model. How can we vary this model to other numbers of units? The key step is the last step of the unit that makes a 45° angle: change this angle and

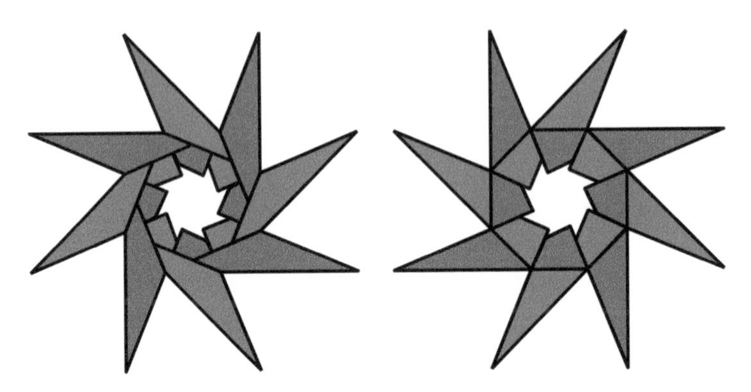

Figure 4.7: *Poly Diag Star 8* (p. 55) made with units omitting a fold.

we can assemble a different number of units. In fact the angle we are interested in is 90°, the internal angle of a square.

For six units we want an internal angle of 120°. *Star 6, Square From Silver Rectangle* (p. 48) shows one way of achieving this and uses a template. *Star 7, Square From Silver Rectangle* (p. 50) changes the angle to approximate 360°÷7 (figure 4.6, A). We could use a template to make a five-piece version. The template is a silver rectangle with the opposite corners folded together: the obtuse angle approximates the interior angle of a regular pentagon. However, the difference in length between the two crease lines on the leftover rectangle is no longer negligible. Although the units are stable when joined, there is extra paper at the reflex angles. Instead, we can use the approximation of a 1:3 rectangle containing a square and 1:2 rectangle (figure 3.15, p. 25). *Star 5, 3 by 2 from Square* (p. 51) finds a 3:2 rectangle inside a 4 by 4 square. *Star 5, 3 by 2 from Square Blintz* (p. 52) is a variation using a 3 by 3 square.

This achieves our aim of varying the number of units from four to five, six and seven. However, the arms of the variation all have the same 45° angle. How can we make this angle relate to the number of units? *Poly Diag Star 6* (p. 53) has arms with 30° at their tips. Varying this makes the eight- and six-piece versions with angles at the arm tips of $180° \div n$ (figure 4.6, B). The oblongs used for *Poly Diag Star 5* (p. 54), *Poly Diag Star 6* (p. 53) and *Poly Diag Star 8* (p. 55) have some interesting properties (p. 25). For eight units, the layers can be thick. To avoid this, leave out the double mountain fold and create a new pattern (figure 4.7).

We have not yet made a three-piece model. So far, the variations of *Star 4, Square From Silver Rectangle* (p. 46) all have one flap aligned with the diagonal of the square. *Star 3, Square From Silver Rectangle* (p. 73) makes this flap stick out so that we can achieve an internal angle of 60° (figure 4.6, C).

This leads us to another way of using 30° and 60° in *Star 6, Bronze Rectangle from Square* (p. 77). Here, the 45° fold of *Star 3, Square From Silver Rectangle* (p. 73) has been changed to 60° (figure 4.6, D). The arms of this star are not skew: the outline forms a kaleidoscope star, a hexagram. *Star 6, Equilateral Triangle from Rectangle* (p. 87) uses a similar method.

Can we make a four-pointed star that also has an outline of a kaleidoscope star? *Star 4, Kite From Silver Rectangle* (p. 79) shows one way: instead of folding on the edge of the square, fold at an angle so that the arm of the star is symmetrical (figure 4.6, E).

Poly Diag Star 5 (p. 54) showed another way of making a five-piece model.

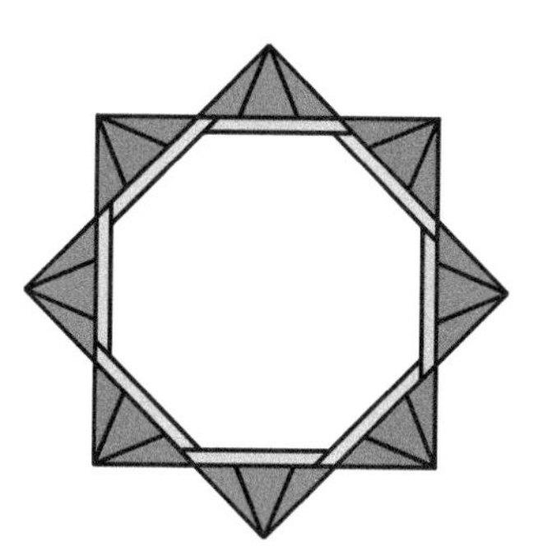

Figure 4.8: *Boat Unit 8 Octagram* made from eight squares. It is generalised from *Boat Unit 7 Octagram* (p. 58), *Boat Unit 6 Hexagram* (p. 57) and *Boat Unit 5 Pentagram* (p. 56).

When *Star 4, Square From Silver Rectangle* (p. 46) is generalised in this way, it can be seen that it is in fact a special case cut from a 1:2 rectangle. *Octagram Wreath* (p. 62) exploits the layers to make a sturdy eight-piece ring (figure 4.6, F). *Swirl 10, Square From Silver Rectangle* (p. 60) uses the same idea on the silver rectangle in the leftover rectangle to approximate the angles of a regular decagon. *Star 5 from Half Silver Rectangle* (p. 80) stretches the starting point to make a hook and lock assembly possible: this time, however, there is no hollow centre.

4.3.3 Sliders

Another way to vary *Star 4, Square From Silver Rectangle* (p. 46) is to narrow the leftover rectangle so that the units of the model can slide more (*Star 4 Slider, Square Silver From Rectangle* (p. 93)). I found this sliding action by accident. The classic slider is Robert Neale's *Pinwheel-Ring-Pinwheel*, also known as *Magic Star* [Neale and Hull 94].

How else can we make units slide? We want the the action to be smooth: this means the sliding tab needs to be free to slide along the runner, but not so free that it comes apart. Ideally the start and end positions are blocked to prevent the units disassembling. *Square Pointer Slider* (p. 94) shows one way. The other sliders experiment with different shapes, and some even have an odd number of units, i.e. *Slider N 5* (p. 103). Notice how *Octagram Slider* (p. 98) makes a slider from *Octagram Wreath* (p. 62) by folding less than a half of the square down.

4.3.4 Overlapped and interlaced regular polygons

Overlapped regular polygons can be interpreted as being interlaced. It is possible to make interlaced squares and regular triangles, or square and triangular frames. *Intersecting Squares* (p. 89) can be seen as a pair of squares that intersect each other. *Star 6 Bronze Wrap* (p. 82) and one side of *Star 6, Equilateral Triangle from Rectangle* (p. 87) can be seen as pairs of interlaced equilateral triangles. *Boat Unit 6 Hexagram* (p. 57) can be seen as a pair of interlaced equilateral triangular frames. This developed from making the *WXYZ* (p. 109) unit from a bronze rectangle. Figure 4.8 shows the *Boat Unit 8 Octagram*, which is not diagrammed in full. It is made from eight squares and is stable but not strong as the flaps are thin and the overlap is small.

4.4 Summary

To summarise, there are two main choices for varying and generalising a flat star. The first is to change the proportions of the paper, which, in turn, alters the lengths and angles of the units. The second is to change a key angle.

For both approaches you will need to know how to divide lengths and angles into equal parts. Here is a summary on how to do this, drawing on some of the folding methods used to make special rectangles (p. 15).

4.4.1 Dividing an edge into equal parts

It is easy to divide an edge into half, quarters, eighths, etc.: less easy when the number of equal parts is odd.

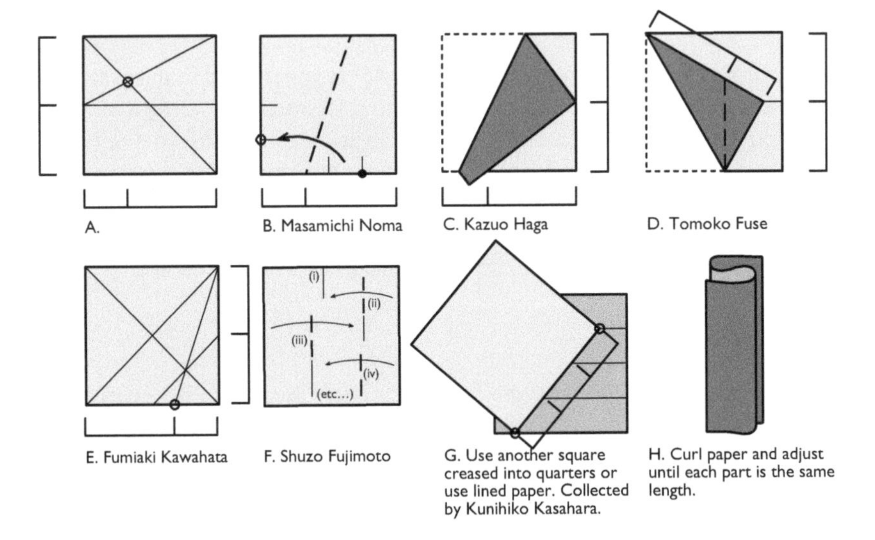

Figure 4.9: Eight methods for dividing the edge of a square into three equal parts.

Figure 4.9 shows a number of ways of dividing an edge of a square into thirds. The best method depends on your objectives. In general we prefer methods that minimise unnecessary creases, tolerate error in folding and use the fewest number of steps. To minimise unnecessary creases, use partial creases; e.g. fold the ends of the crease line but do not press down on the middle part. Note that some models will use some of the construction creases. Some of the methods rely on properties of the square but others will generalise to oblongs. Some methods generalise from thirds to other numbers of equal divisions.

With practice, curling the paper (method H) to divide an edge into thirds can be quick and leave few extraneous creases on the paper. Method F (Fujimoto) works well for thirds, fifths and other numbers that are not a power of 2. Method G (equally spaced parallel lines) is accurate and leaves no unnecessary creases.

If you need to make the same divisions on a number of sheets, consider using a template (figure 4.10).

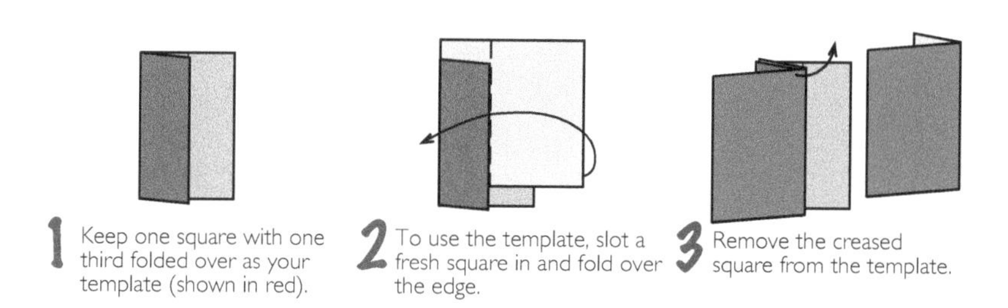

Figure 4.10: Using a template to speed up folding thirds.

4.4.2 Dividing a right angle into equal parts

As with dividing an edge into equal parts, dividing an angle into equal parts is easy when the number of parts is a power of two: 90°, 45° and 22.5° are commonly used angles in origami. These angles let us make regular octagons.

The section on bronze rectangles discussed how to trisect a right angle to obtain 30° and 60° (figure 3.9, p. 20). These angles let us make equilateral triangles and regular hexagons.

More difficult is dividing a right angle into five equal parts, which is needed for regular pentagons. In practice, approximations are usually good enough. We have already seen that silver rectangles (p. 18) and 1:3 rectangles (p. 24) are used for this. One way to improve accuracy is to use Fujimoto's iterative method: it works for angles as well as lengths.

Figure 4.11 shows a different method to improve the 1:3 approximation for a unit square. *Pentagram* (p. 76) uses this method as the extra accuracy is essential for the small locking flap. First crease the diagonal. Then on the left edge of the square, pinch the midpoint M and the quarter point Q. Bisect the 45° angle in the lower right corner, pinching K. Then pinch H, halving the distance from K to the top left corner. Optionally pinch the midpoint of QH, or locate it by eye. The fold to make goes from just below this point to the lower right corner. At the same time, the midpoint M will land just above the diagonal.

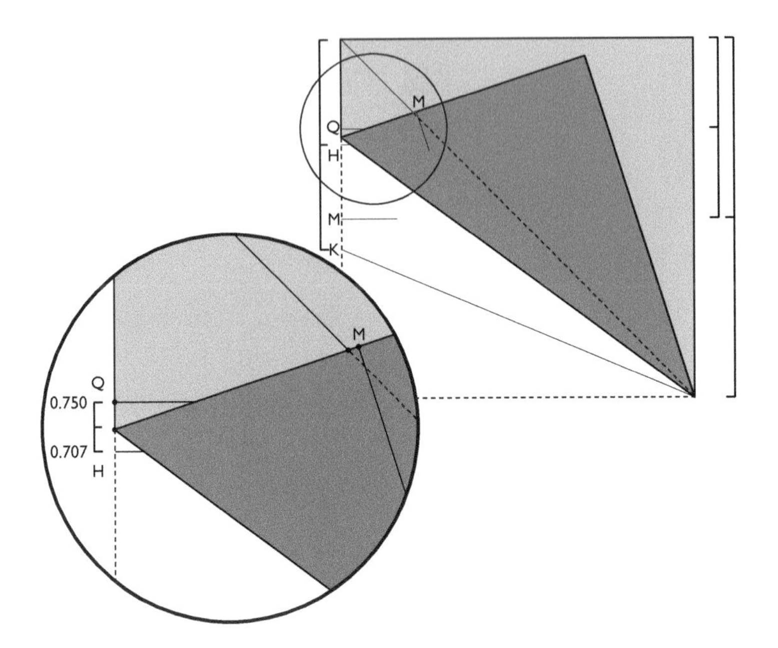

Figure 4.11: Improved method for approximating 36° in a unit square.

Dividing an angle into seven equal parts by folding alone is not easy. To approximate a regular heptagon, we want to make the central angle 360°÷7 = 51.429°. Figure 3.13 (p. 24) showed one simple approximation which lets us approximate a regular heptagon. *Silver Star 7* (p. 74) uses a different method to make 52.897°. *Boat Unit 7 Heptagram* (p. 58) uses a much better approximation: tan(5÷4) = 51.340°, which is close to 360°÷7 = 51.429°.

4.4.3 Dividing 180° and 360° into equal parts

It is straightforward to extend the methods for dividing a right angle into equal parts to divide a half turn and a full turn into equal parts. Figure 4.12 shows a couple of special cases. Both are squares book-folded in half so that the centre of the square can be divided into either tenths or thirds. *Blintz Icosidodecahedron* (p. 111) locates a silver rectangle using angle bisectors of 45° and 90°. *WXYZ* (p. 109) trisects 180° using a $\frac{3}{4}$ line.

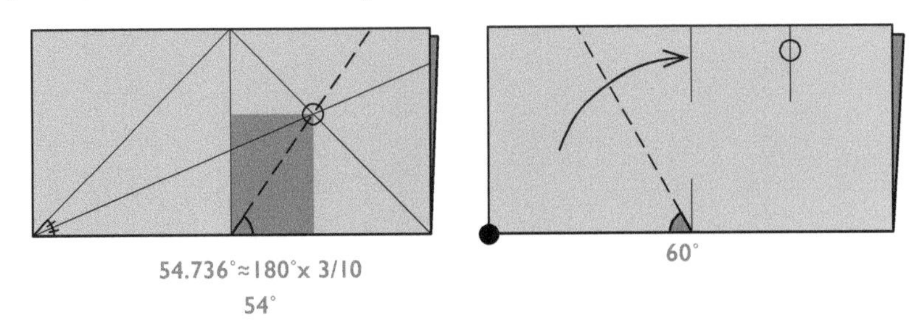

Figure 4.12: Left: locating a silver rectangle approximates 54°. Right: trisecting 180° using a $\frac{3}{4}$ line.

Chapter 5

Tips on Paper and Folding

5.1 Paper

You should be able to make all of the models in this book using commonly available paper. Ordinary 80 gsm A4 photocopy paper is readily available in a variety of colours. For squares you can cut your own squares from oblongs, use paper from memo cubes or 15 cm origami paper (figure 5.1, left).

In general, choose papers that are strong (will not rip easily), not too thick (makes thick layers and causes inaccuracy) and will take and hold a crease well. This means you should normally avoid newsprint, paper-backed foil and card stock. However, if you want to make large models then you might need to use slightly thicker paper.

Reusing paper like junk mail and flyers is fine, especially for practice, but be careful if the paper has personal data. You can fold paper from magazines but this can be unpleasant as the coating rubs and cracks off the coated paper. For fine work, choose acid-free paper that will not discolour with age.

5.2 Paper colour and patterns

Commercially produced origami paper usually has different colours on each side: this can help you follow diagrams as the two sides are clearly differentiated. It can improve the appearance of some finished models, e.g. *Octagram Paper Cup - Kite Wheel* (p. 63). Other times paper that is the same colour on both sides is easier to obtain and works better for some models, e.g. it can prevent the white side of the paper showing in *WXYZ* (p. 109).

You can use plain paper for all of the models in this book. Patterned paper can enhance some models if you choose suitable paper (figure 5.1, right). Avoid busy patterns that can overwhelm and distract the effect of the finished model.

Usually you use different colours for units that are next to each other. For models with an even number of units, using two different colours is convenient and attractive. For prime numbers like five and seven, you can use all different colours. If you do not have a sufficient number of colours, a single colour can be effective.

DOI: 10.1201/9781003184492-5

Figure 5.1: Left: a selection of paper for folding. A4 printer paper and memo cube paper is usually 80 gsm and the same colour on both sides. Commercial origami paper is usually white on one side and printed with a colour or pattern on the other. Common sizes are 75 mm and 15 cm squares. An envelope opener is safe and convenient for cutting paper. Right: *Pinwheel Square Slider* (p. 95) and *Handshake Wreath 8* (p. 66) folded from patterned paper. Which patterns enhance the final appearance? Which patterns are distracting?

5.3 Paper size

Convenient paper sizes are suggested for each model so that the results fit your hands. You can usually use squares or rectangles at least 75 mm wide. If you want to know the size of the result for other paper sizes, use the finished model ratio given in the diagrams. The ratio is usually the longest length in the largest configuration. If the starting rectangle is not square, the shortest length is assumed to be 1.

5.4 Preparing paper

Most of the models use rectangles of convenient proportions. Sometimes you will need to cut paper.

For cutting, certain kinds of envelope openers are convenient to use. They are small, portable, have no exposed blade and are relatively easy to use. You fold the paper where you want to cut and then slide the envelope opener along the folded edge. However, they work best for cuts perpendicular to a raw edge and are not so good for other angles, e.g. 60°. Remember not to crease too hard or the blade will not catch properly.

5.5 Folding the modules

You can fold paper almost anywhere, but if you have a choice then use a bright and comfortable place for folding. Make sure you have enough suitable paper and have clean hands before you begin. Paper absorbs moisture and oil from your fingers so try to handle the paper as little as possible.

Like most modular origami, sharp accurate folding will make assembly easier. Fold on a hard surface and be careful to keep edges or location marks lined up as you fold.

As you fold more units, you should find your accuracy improves as your "muscle memory" learns the steps and you may find better ways to make the units than the diagrams suggest. This is why you should normally avoid an "assembly line" method of folding such as step 1 on each sheet of paper, then step 2 on each sheet, then step 3 on each sheet, etc. Not only is this dull and unsatisfying compared with making complete modules one at a time, you might find that you've made the same mistake on all of the modules!

5.6 Assembling the modules

Once you have two modules try joining them together as this makes folding the modules more motivating. It also lets you find out if you need to make any minor adjustments to the modules, perhaps because of the size and thickness of your paper. Usually it is better to keep adding modules to the same model. Do not make separate sub-assemblies and then join them together—this can become confusing and is usually inefficient.

2D models are mostly straightforward to assemble. Joining the last module can be hard at first. You might need to gently bend the model or 'flip' the first modules forward to access them. This means lifting the first module and pulling it towards you, otherwise you will make a helix that could go on forever. Steps 9 and 10 of *Star 4, Square From Silver Rectangle* (p. 46) show this in detail. When you pull the last module towards you, there are two main choices of where the final module goes. The final module goes above the first module if each module is on top of the next like *Star 4, Square From Silver Rectangle* (p. 46), or inside the first module if each module is inserted into the next as in *Pinwheel Square Slider* (p. 95). Put simply, join the first and last units in the same way as all of the other units are joined.

Sometimes it is useful to remove the first unit and use it as a new "last" unit. You know that the "last" unit joins well with the "first" unit, i.e. the old second unit. This is useful where a locking fold is made during assembly, e.g. *Swirl 10, Square From Silver Rectangle* (p. 60).

3D models are usually harder to assemble than 2D models. If the last module is hard to add, loosen the other modules and try again.

For difficult and awkward assemblies, you might need some tools. The first type of tool lets you poke and tuck flaps in difficult places—cocktail sticks and nail files work well. The second type of tool temporarily holds modules together—you can use paperclips, mini clothes pegs or pressure-sensitive masking tape depending on the model. Paperclips can leave marks so do not leave in place for too long. Also do not leave masking tape on for too long either as it might bond to the paper and rip when removed.

5.7 Following the instructions

Diagrams are like a comic strip that you follow, performing the folds in each step. Do not let the origami terminology intimidate you: you can usually tell what's required by looking ahead to the next step. You can turn the paper to make it easier to fold—the paper does not have to be in the same orientation as the diagram. For example, to make a mountain fold you might find it easier to turn the paper over, valley fold and then turn over again. Or a vertical valley fold is easier to make if you rotate the paper 90° so that you are folding away from your body.

If you are a beginner then start with the simpler models first. Models earlier in the book have more detailed instructions. Later models tend to be briefer, not only because it saves space, but also because as you become more experienced, you can work with larger "chunks" of information. In general the simpler models are at the beginning of each group of models. Also note that some models have different difficulties of folding the modules and assembling them. For example, the modules of *Star Ball* (p. 107) are simple to fold but are not so simple to assemble.

Beginners usually appreciate the detailed text in each step, especially for assembling the modules. Many experienced folders skip the words in each step: besides fluency, one reason is that they often fold from diagrams written in languages that they cannot read, so manage without reading the words. Even so, the words can still help you by using your verbal communication channel and working memory. Sometimes the words give helpful tips that you might miss if you do not read the diagram carefully, for example only fold between the existing crease lines.

Note that diagrams are usually drawn with the layers of paper slightly spread out, even if they are meant to align exactly. This helps you understand how the layers are arranged. If you are meant to leave gaps, this will be specifically highlighted. As mentioned elsewhere, diagrams use different colours to distinguish the two sides of the paper. However, for most models you can use paper that is the same colour on both sides: look at the final model to decide which kind of paper you should use.

If you are stuck, study the diagram carefully: is there a line or arrow that you've missed? Do the flaps and layers of paper correspond to the paper in your hand? If taking a break and trying again doesn't work, ask someone else for help. It doesn't have to be an origami expert as a fresh pair of eyes might be all that's needed. If nobody is able to help, try starting again from the beginning, carefully following every symbol and instruction. Despite the best intentions of all involved, diagrams may have mistakes so see if you can work around the problem: can you carry on even though you've missed out a step?

See Symbols and Procedures for more hints on understanding diagrams.

Chapter 6

Symbols and Procedures

The diagrams in this book use symbols based on standards developed by Akira Yoshizawa, Robert Harbin, Samuel Randlett and others. If you take some time to understand these symbols then they will soon become second nature. Not only will this save you time and make following diagrams more pleasant, you will also be able to fold from diagrams in languages that you cannot read. You will also develop the valuable skill of folding imaginary paper in your mind.

At each step you can usually look ahead to see the result that you are aiming for. However, sometimes you will need to look carefully at the symbols. Most often you will need to distinguish valley (concave) and mountain (convex) folds: the former uses a dashed line and a plain arrowhead; the latter uses a line of dashes and dots with a hollow half arrowhead. See overleaf.

The standard symbols have some quirks. The turn over symbol might evoke turning over the model, but taken literally would turn the model over again. The cut symbol uses scissors but using a knife is better in practice — the scissors are a familiar design trope for cutting coupons out of magazines, etc.

As the diagrams in this book use different shading for the two sides of the paper, most diagrams omit the symbol for coloured side up / down. Repeat symbols are rarely used as it's usually obvious where you need to repeat the steps — read the text for specific information about repeats. For a similar reason, the enlarged view symbol is not always used; and curiously, no standard reduced view symbols exists.

Some common compound folds are given special names: see page 43. This book only refers to a few procedures by name: the squash fold, (inside) reverse fold, outside reverse fold, rabbit ear, preliminary fold and waterbomb base. Most models only use valley and mountain folds with a few squash and reverse folds.

There is some other shorthand: a "blintz fold" means folding all the corners of a square to the centre. A "kite fold" is a square with two adjacent edges folded to a diagonal: this can fly as a working kite when rigged with cord. Informally, a kite fold is the result of bisecting an acute angle of an isosceles right-angled triangle. The meeting point of the 22.5° fold and the edge is known as a "kite point".

You will find two other procedures used in some models: dividing a length into equal parts and folding 60°. It is easy to divide a length into halves, quarters, eighths but thirds and fifths can be also made without a ruler. A good way is Shuzo Fujimoto's iterative method. Folding 60° is a special case of Abe's trisection of an acute angle.

DOI: 10.1201/9781003184492-6

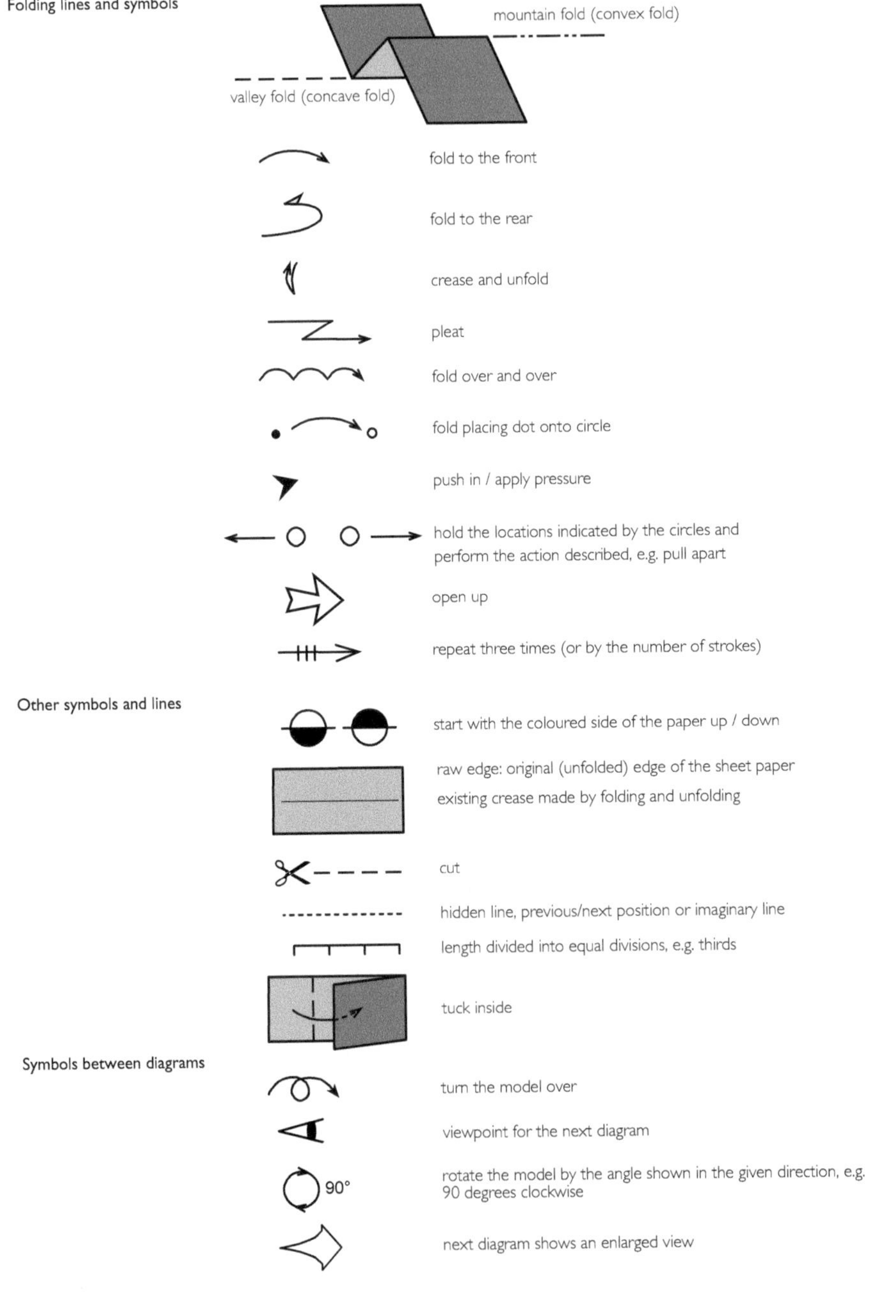
Folding lines and symbols
mountain fold (convex fold)
valley fold (concave fold)
fold to the front
fold to the rear
crease and unfold
pleat
fold over and over
fold placing dot onto circle
push in / apply pressure
hold the locations indicated by the circles and perform the action described, e.g. pull apart
open up
repeat three times (or by the number of strokes)
Other symbols and lines
start with the coloured side of the paper up / down
raw edge: original (unfolded) edge of the sheet paper
existing crease made by folding and unfolding
cut
hidden line, previous/next position or imaginary line
length divided into equal divisions, e.g. thirds
tuck inside
Symbols between diagrams
turn the model over
viewpoint for the next diagram
90°
rotate the model by the angle shown in the given direction, e.g. 90 degrees clockwise
next diagram shows an enlarged view

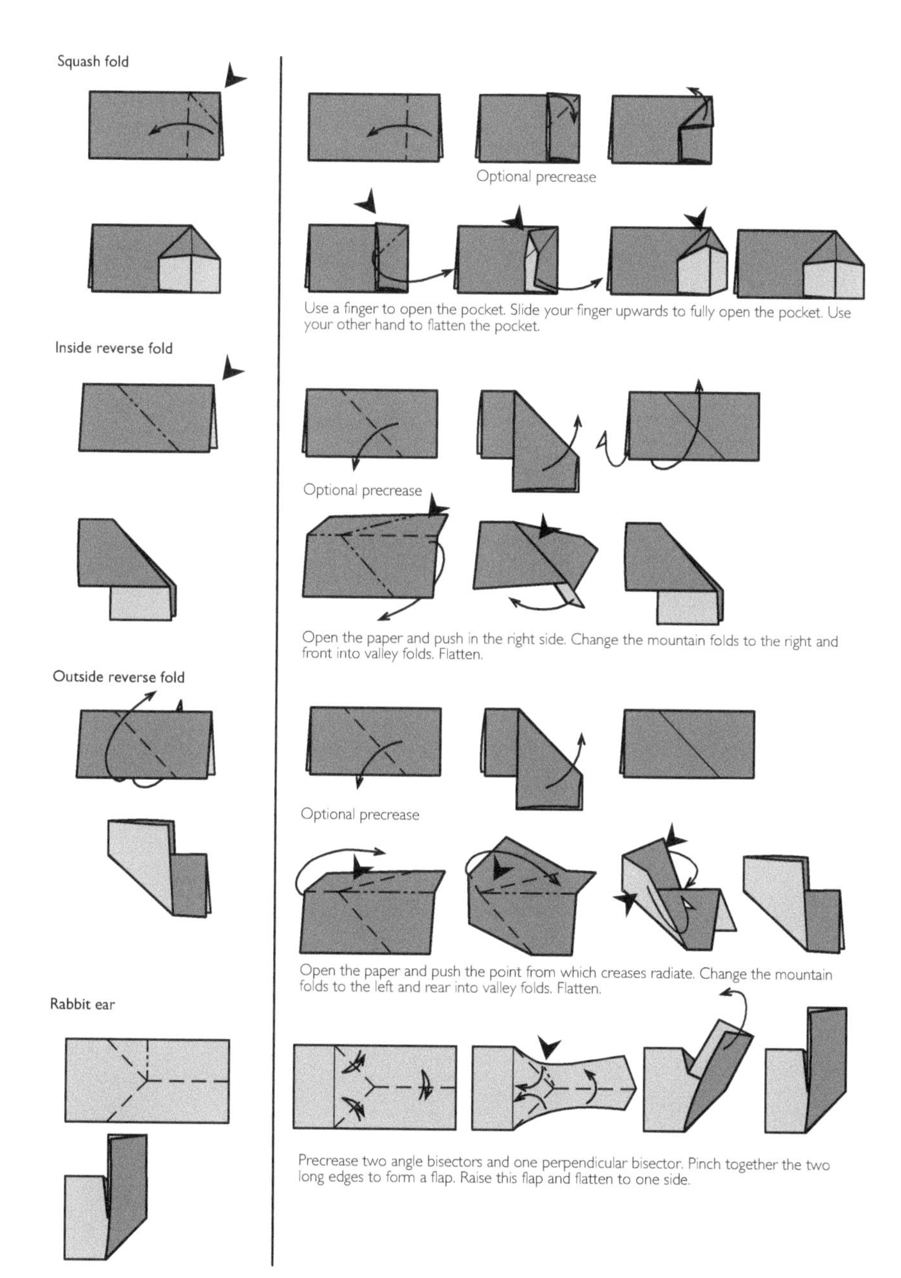
Squash fold
Optional precrease
Use a finger to open the pocket. Slide your finger upwards to fully open the pocket. Use your other hand to flatten the pocket.
Inside reverse fold
Optional precrease
Open the paper and push in the right side. Change the mountain folds to the right and front into valley folds. Flatten.
Outside reverse fold
Optional precrease
Open the paper and push the point from which creases radiate. Change the mountain folds to the left and rear into valley folds. Flatten.
Rabbit ear
Precrease two angle bisectors and one perpendicular bisector. Pinch together the two long edges to form a flap. Raise this flap and flatten to one side.

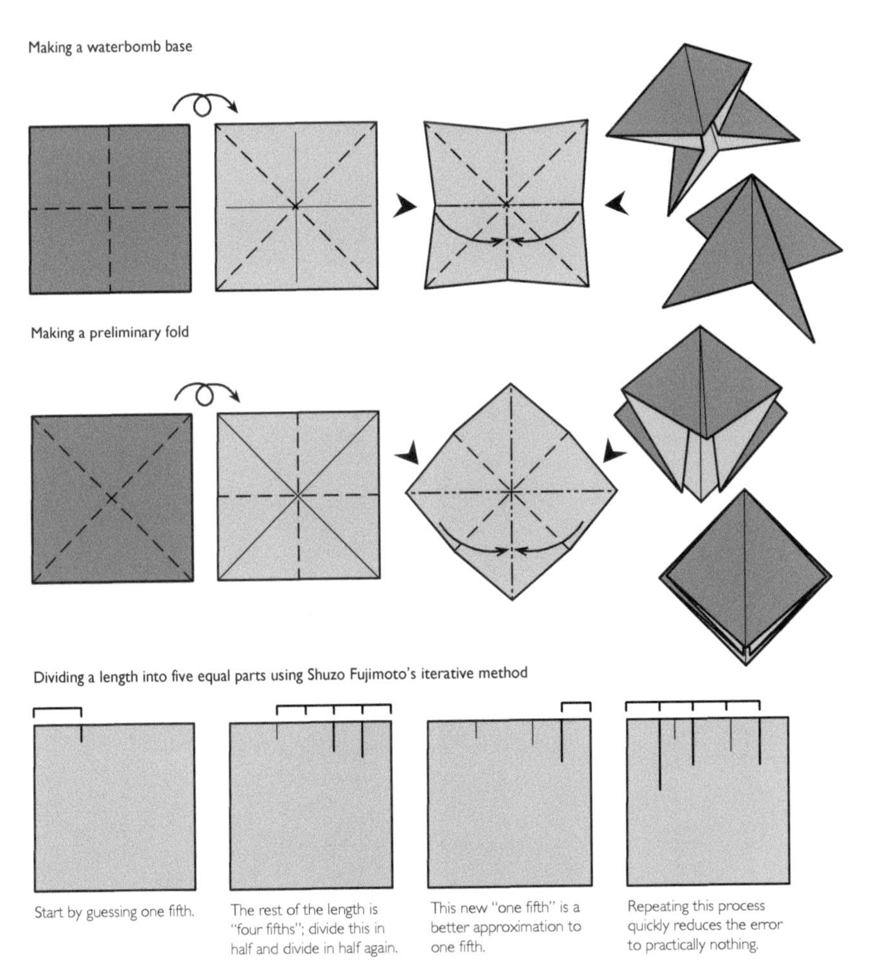

Trisecting a right angle

Fold the lower left corner onto the middle line. At the same time the fold starts at the top left. This fold appears in different guises but is essentially the same fold.

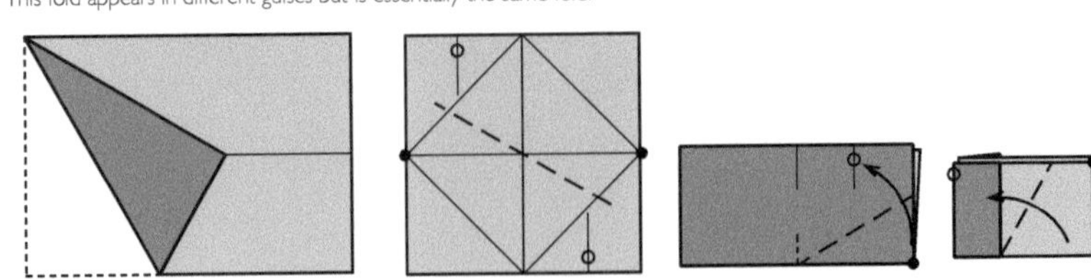

Chapter 7

Ring Stars

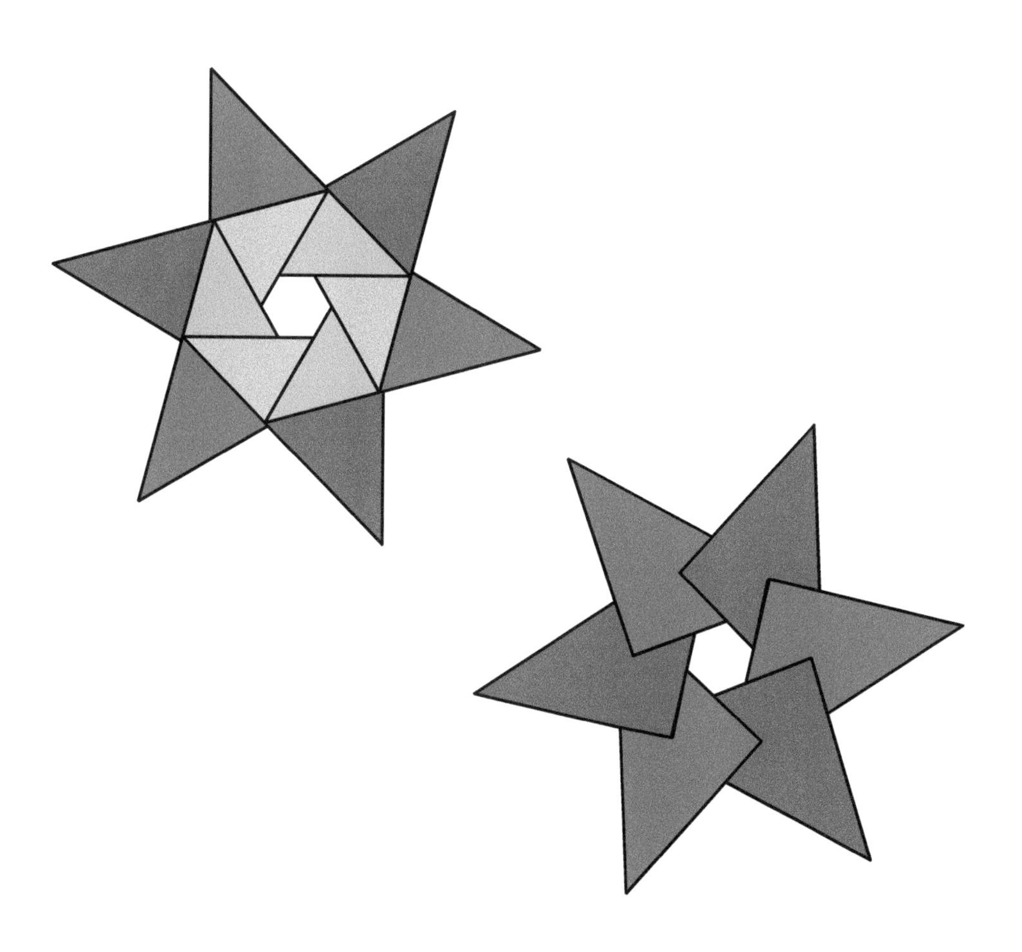

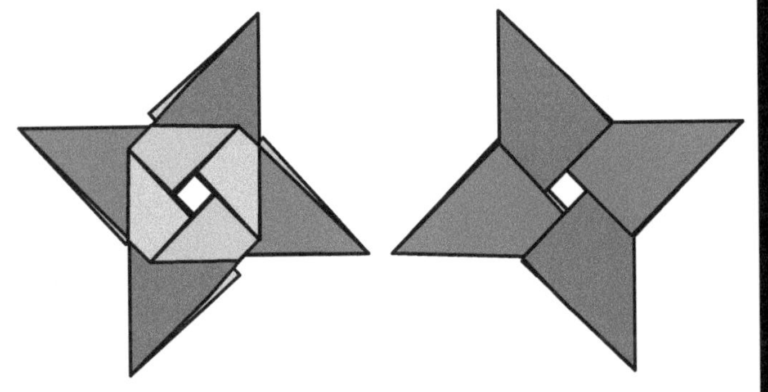

Star 4

Square From Silver Rectangle

★

Use four silver rectangles
or ratio between 1.25:1 and 1.5:1
Finished model ratio 2.3

The first two steps are familiar from making a square from an oblong. However, instead of cutting off the leftover rectangle, two 45° folds make flaps for a secure lock. However, the units can slide a little: when made from A7 rectangles, the units can slide about 15 mm. Note that you can use rectangles other than silver rectangles, as long as they are not longer than 3:2. 8.5 by 11 inch paper cut into quarters would work.

Module ★

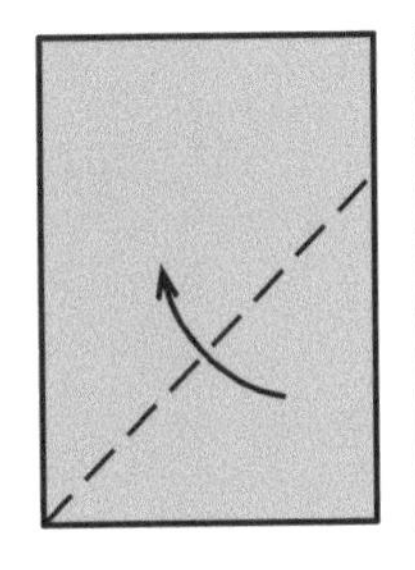

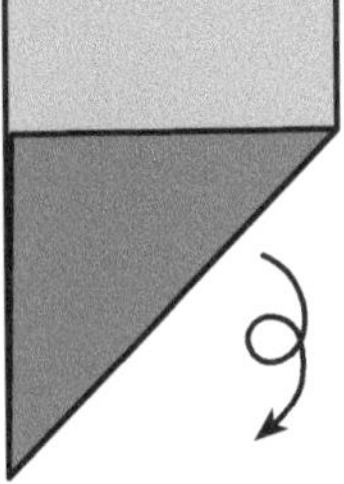

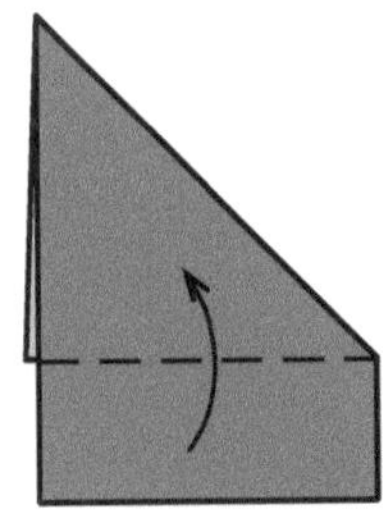

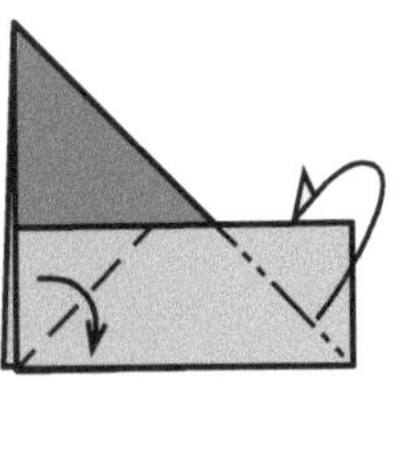

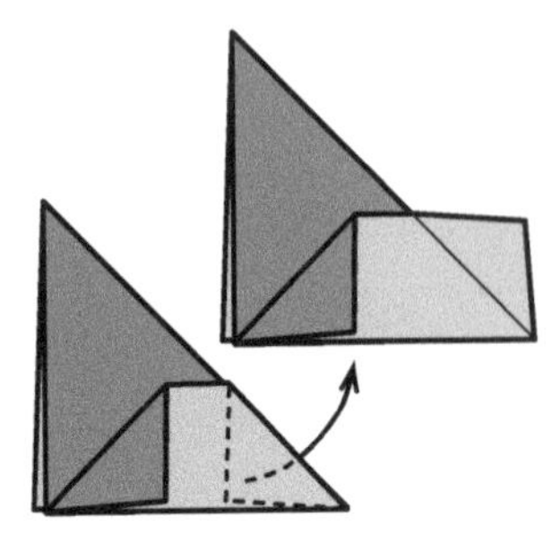

1 Bisect the lower left right angle: valley fold the bottom edge to the left vertical edge. Turn over.

2 Fold the lower flap up so that the fold aligns with the raw edge behind it. Make two 45° folds on the new flap.

3 The mountain fold of the right flap that follows the folded edge behind. Unfold the right flap.

Assembly ★

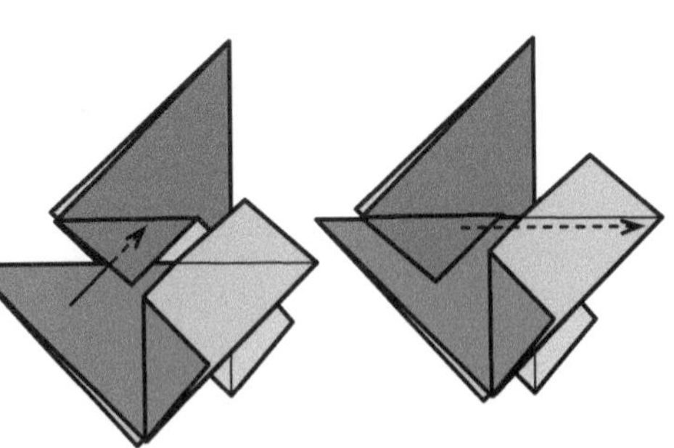

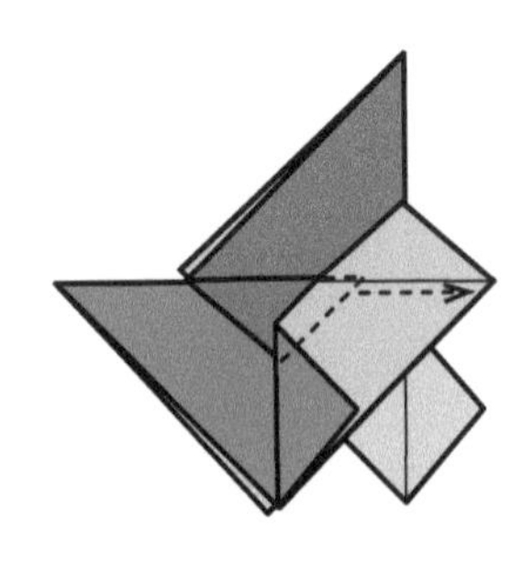

1 Rotate the first unit 45° anticlockwise. and the second unit 45° clockwise. Hook the folded flap of the second unit between the layers of the first unit.

2 After hooking the second unit onto the first, slide the second unit to the right as far as it will go.

3 Slide the second unit to the right as far as it will go.

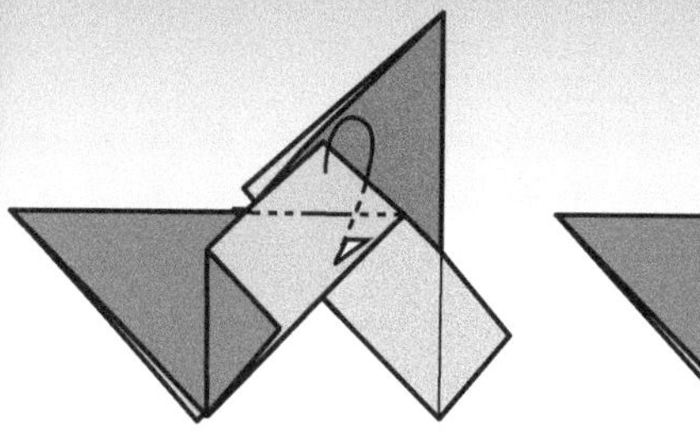

4 Mountain fold the flap of the first unit behind to finish joining the first two units.

5 Two units joined.

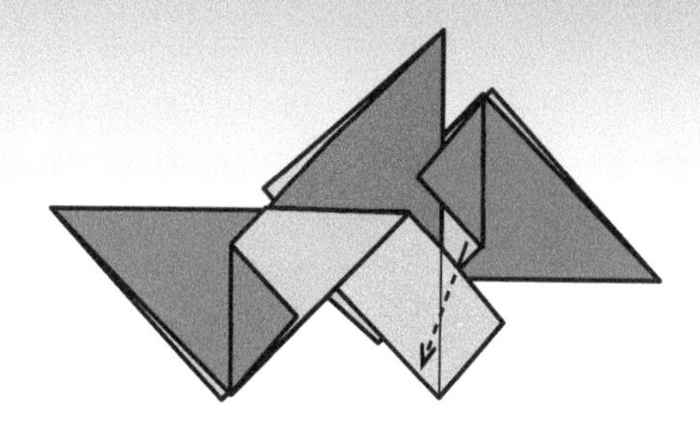

6 Hook the third unit onto the second unit. Slide it down as far as it will go.

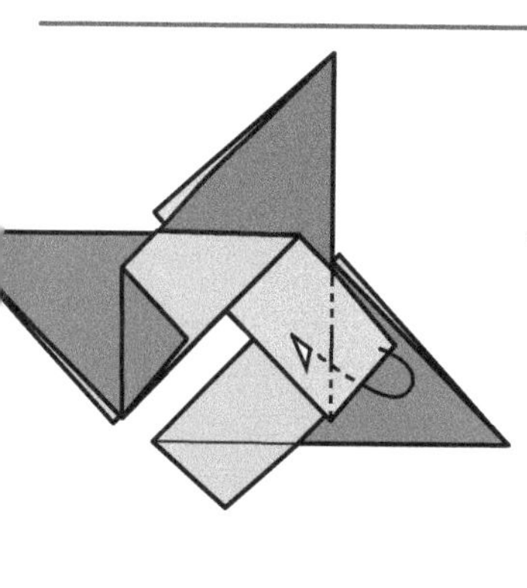

7 Mountain fold the flap to finish joining the third unit to the second.

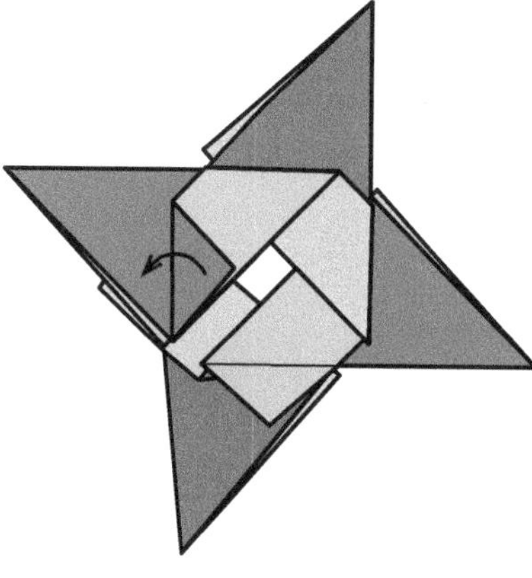

8 Hook the fourth unit onto the third unit. Slide and mountain fold to lock. Unfold the small triangular flap of the first unit.

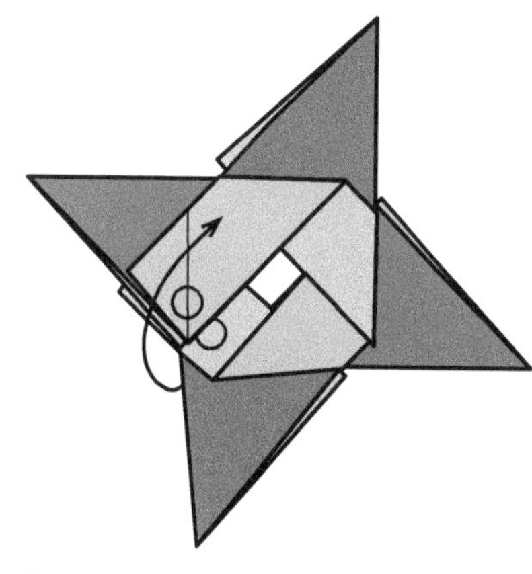

9 To join the fourth to the first, lift the end of the fourth unit up and over the first unit.

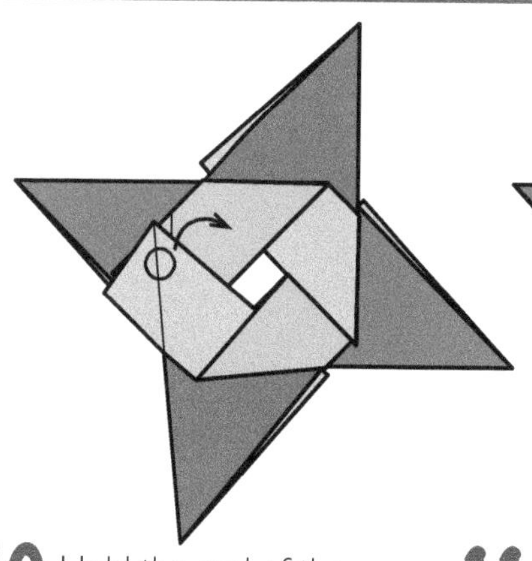

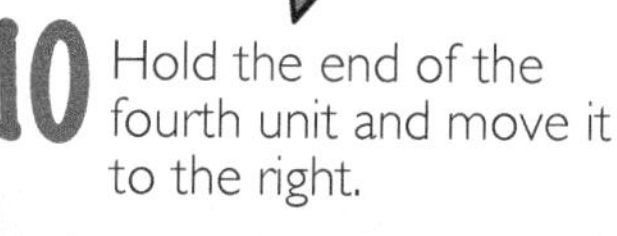

10 Hold the end of the fourth unit and move it to the right.

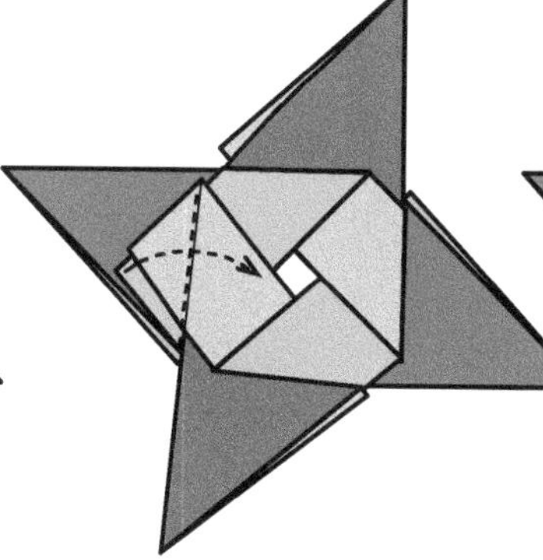

11 Valley fold the small flap of the first unit into the fourth unit.

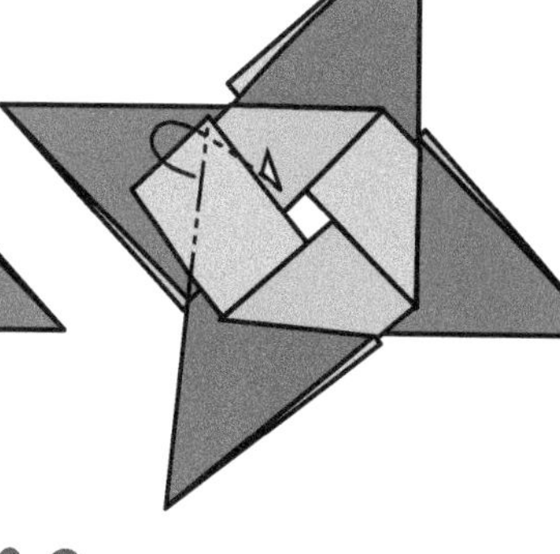

12 Mountain fold the small flap of the fourth unit into the first unit to finish joining all units.

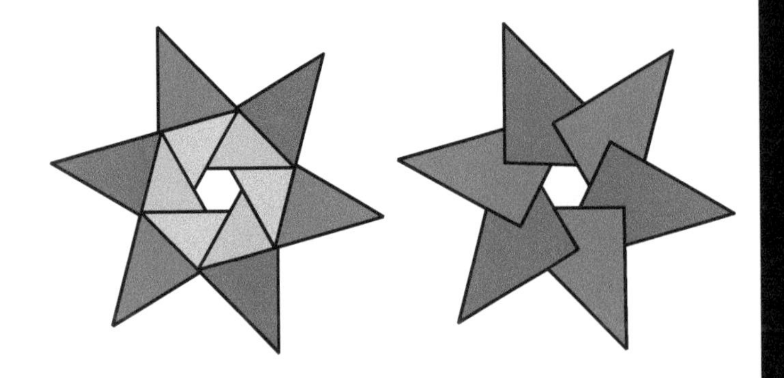

Star 6
Square From Silver Rectangle

★

Use six silver rectangles for the units and one for the template
Finished model ratio 2.5

This is a six-piece variation of the previous model. The 60° fold is accurate but the unit uses an approximation: the length of the fold in module step 5 is slightly longer than the fold in step 2. You can fold the unit without a template but this might leave extraneous creases that spoil the final appearance.

Template ★

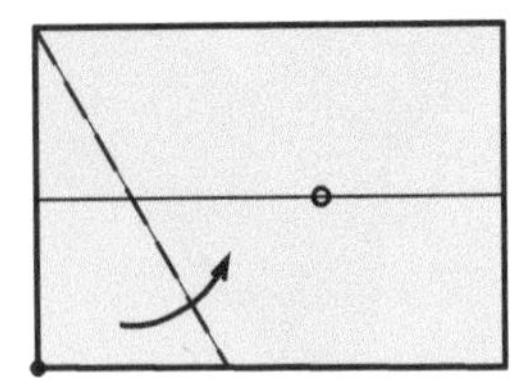

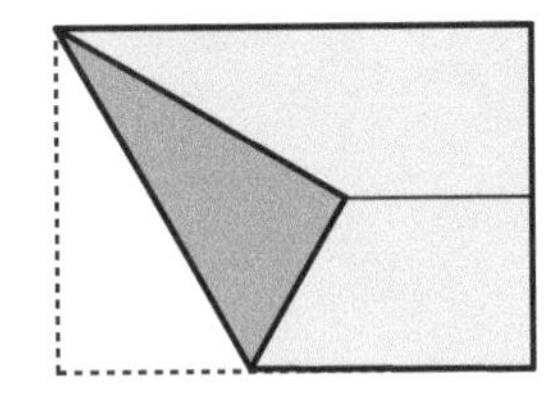

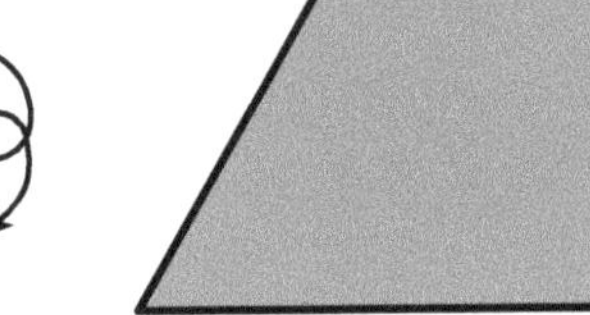

1 Fold in half bringing long edges together, then unfold. Fold the lower left corner onto the central crease line and make the fold go through the top left corner.

2 Turn over.

3 You only need one template.

Module ★

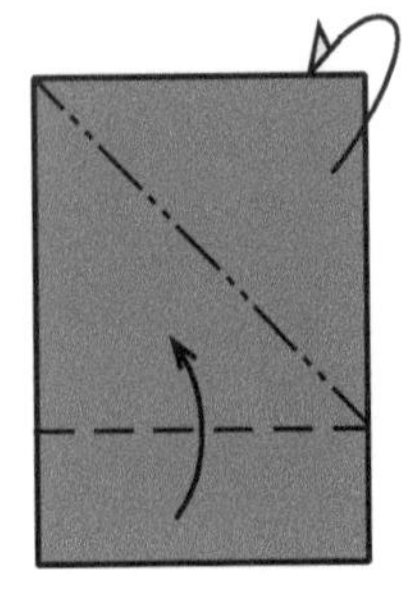

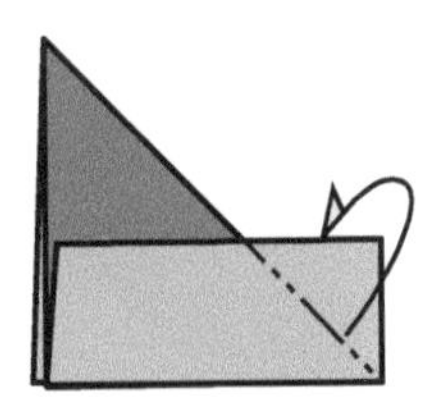

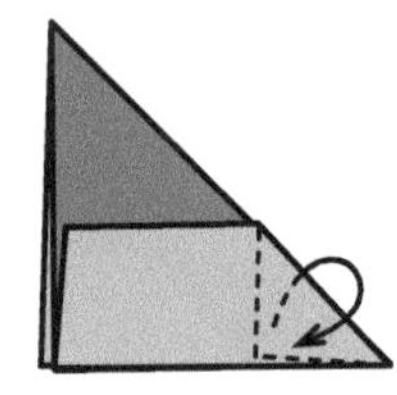

1 Bisect the top left right angle with a mountain fold. Valley fold the bottom edge to the left vertical edge.

2 Fold the right flap behind using the diagonal fold as a guide.

3 Unfold the right flap and valley fold it.

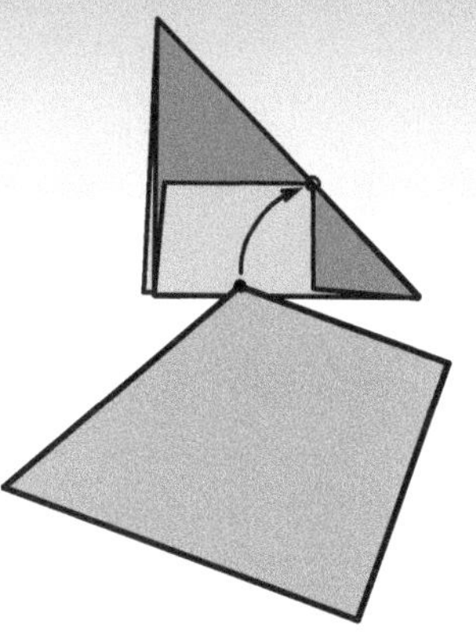

4 Insert the template into the unit. Put the short edge of the template under the small right flap and make the obtuse angled corners meet.

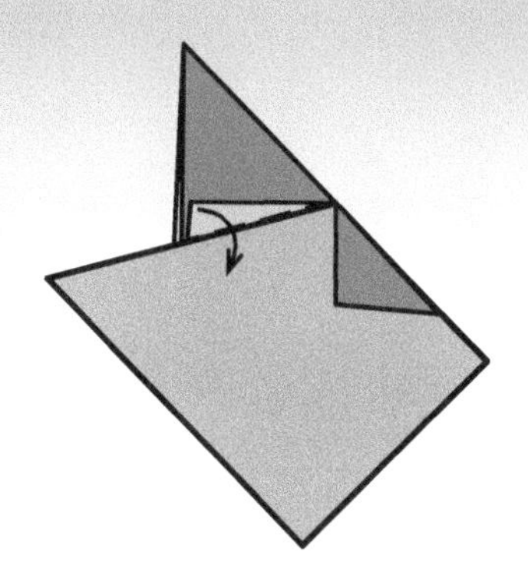

5 Fold the small right angled flap of the unit over the edge of the template.

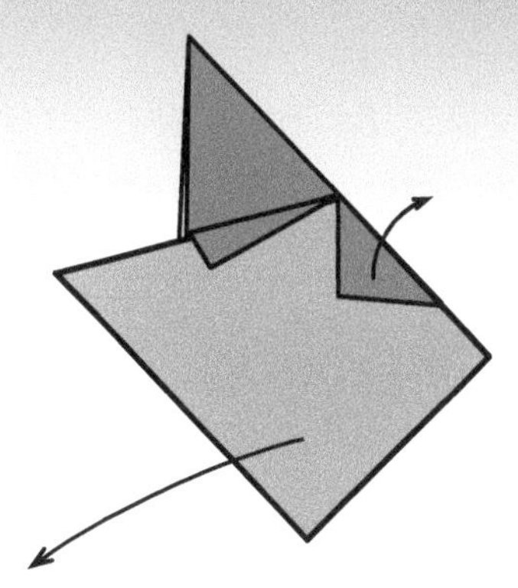

6 Remove the template. Unfold the small right flap. Make a total of six units.

Assembly ★

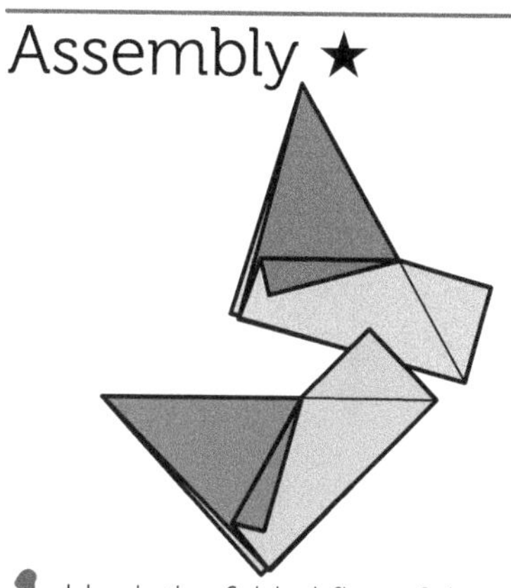

1 Hook the folded flap of the second unit between the layers of the first unit.

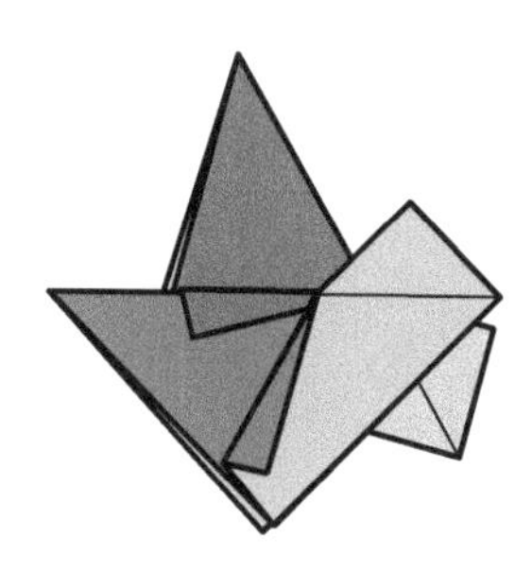

2 After hooking the second unit onto the first, slide the second unit to the right as far as it will go.

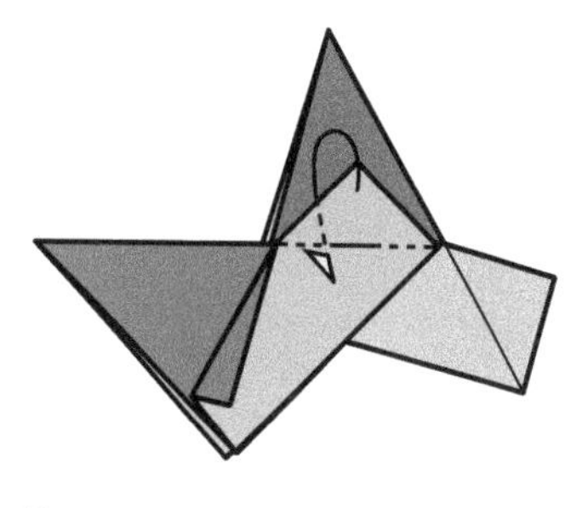

3 Mountain fold the flap of the first unit behind to finish joining the first two units.

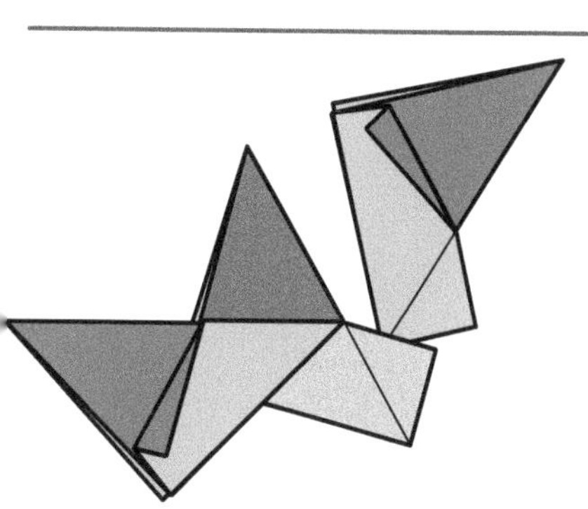

4 Two units joined. Continue adding units in a clockwise direction.

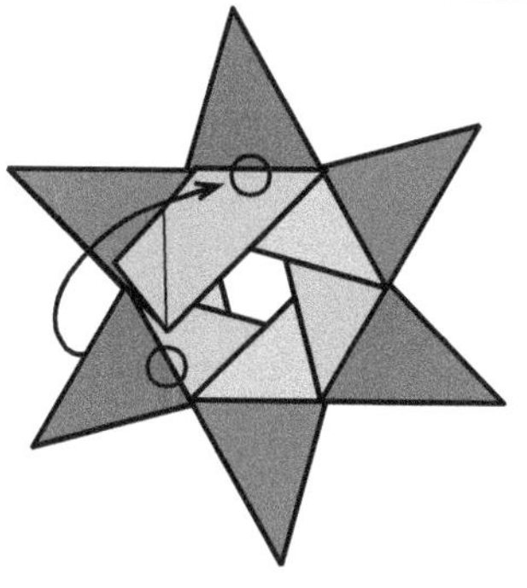

5 To join the last unit to the first, lift the end of the last unit up and over the first unit.

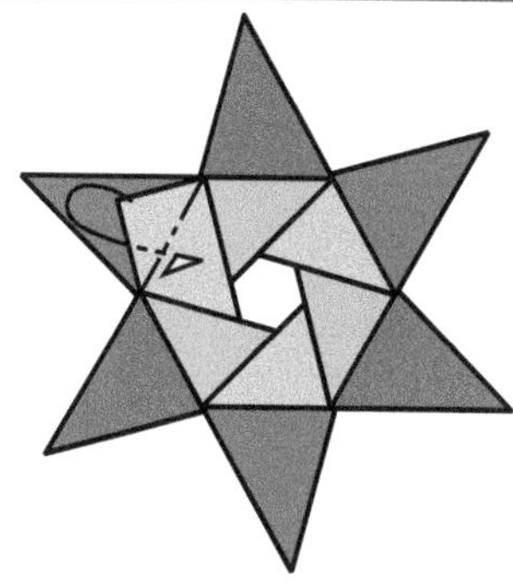

6 Hook the first unit into the last unit. Mountain fold the small flap of the last unit to finish the model.

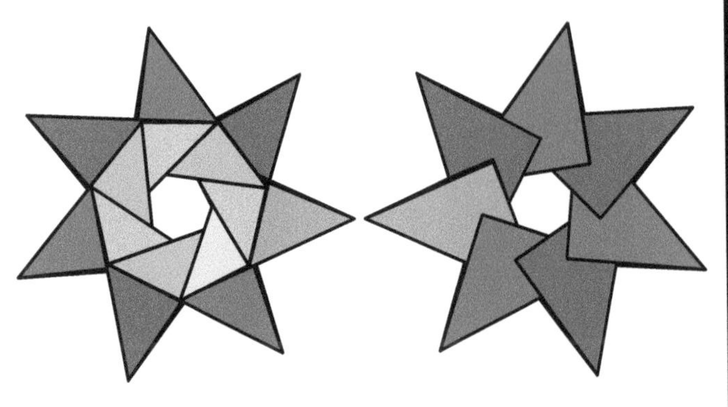

Star 7

Square From Silver Rectangle

★

Use seven silver rectangles like A7
Finished model ratio 2.5

This seven-piece model follows on from the previous models. It approximates the interior angle of a regular heptagon (128.57°) with 126.184°. This comes from taking a quarter of $\tan^{-1}(1/\sqrt{2})$ away from 90° and then adding 45°.

Module ★

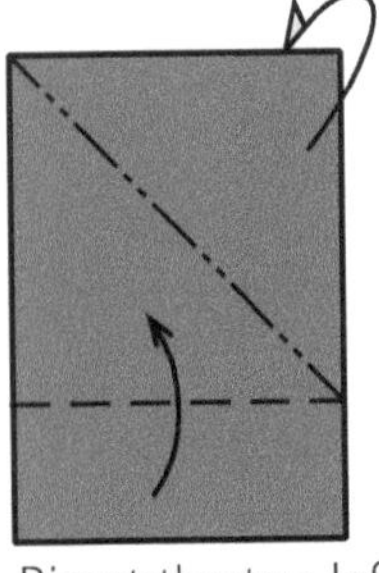

1 Bisect the top left right angle with a mountain fold. Valley fold the bottom edge to the left vertical edge.

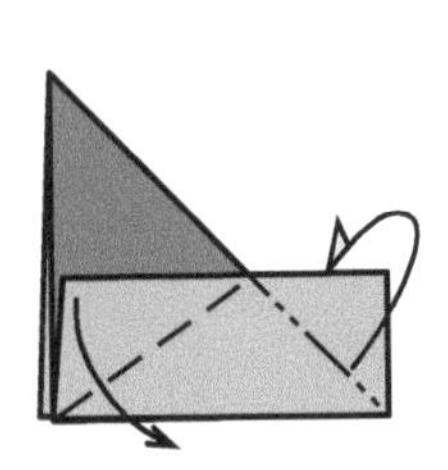

2 Fold the right flap behind. Fold the diagonal of the small silver rectangle.

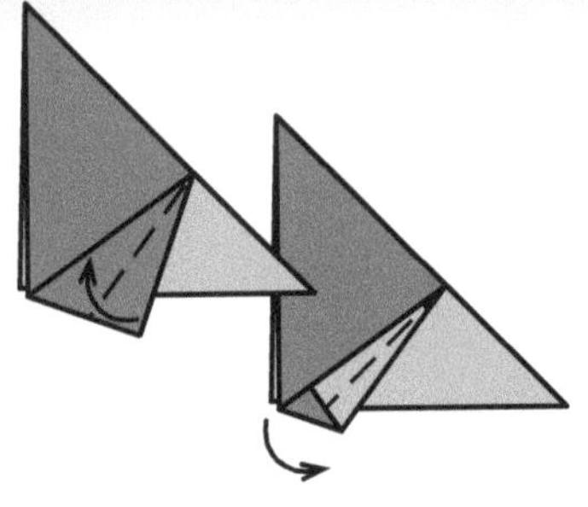

3 Fold the raw lower edge of the flap up to bisect the angle. Then fold the edge down to bisect the new smaller flap.

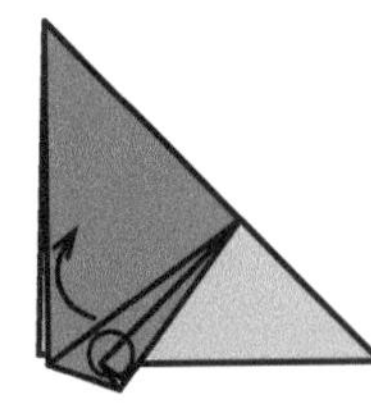

4 Hold the corner and pull away, keeping the most recent fold in place. Unfold the right flap.

Assembly ★

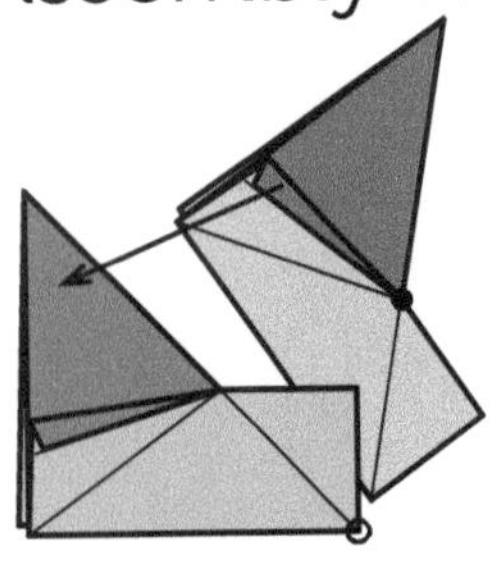

1 Hook the folded flap of the second unit between the layers of the first unit.

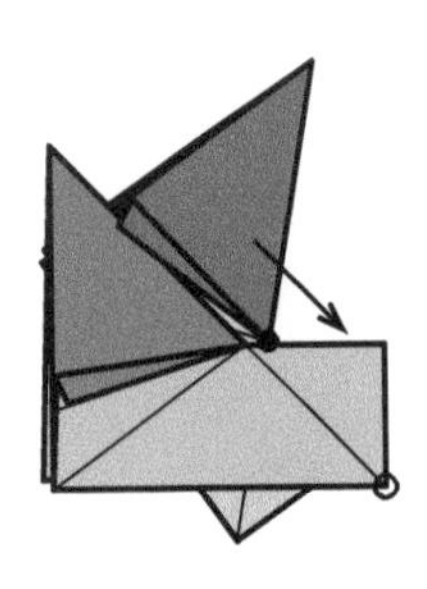

2 After hooking the second unit onto the first, slide the second unit to the right as far as it will go.

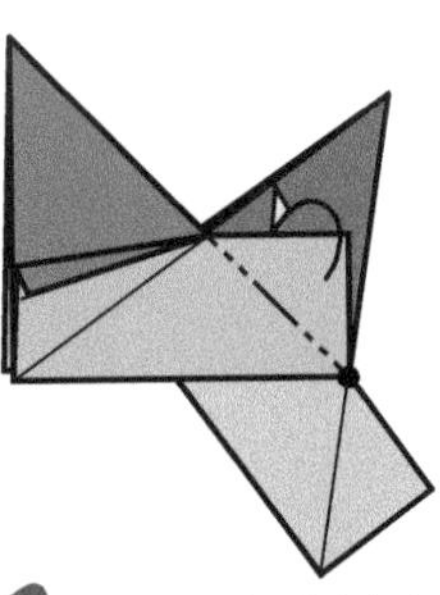

3 Mountain fold the flap of the first unit behind to finish joining the first two units.

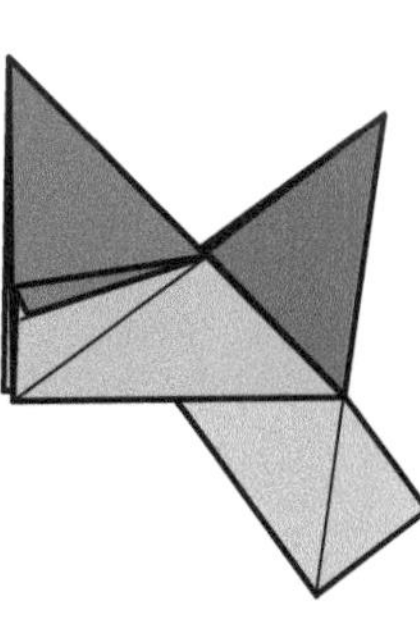

4 Continue adding units in a clockwise direction. To join the last unit to the first, first lift the end of the last unit up and over the first unit.

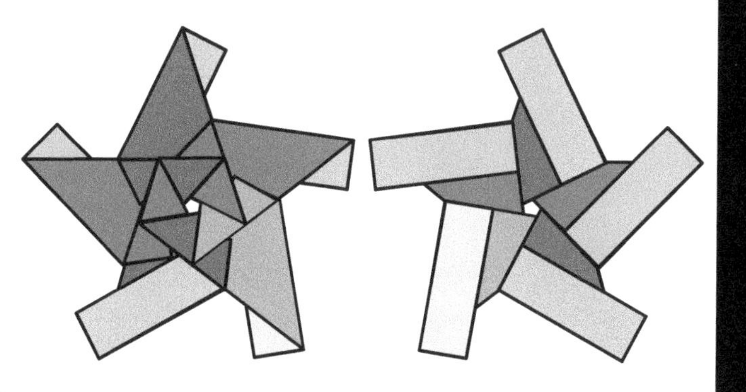

Star 5

3 by 4 Rectangle From Square

★★

Use five squares about 100 mm
Finished model ratio 1.7

A 4 by 4 square contains a 3:4 rectangle. The excess paper could be folded away but is left alone in this version. This model makes a good pinwheel when a pencil is put through the hole.

Module ★★

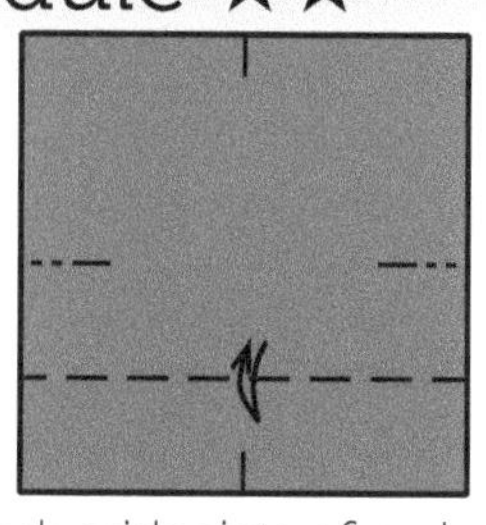

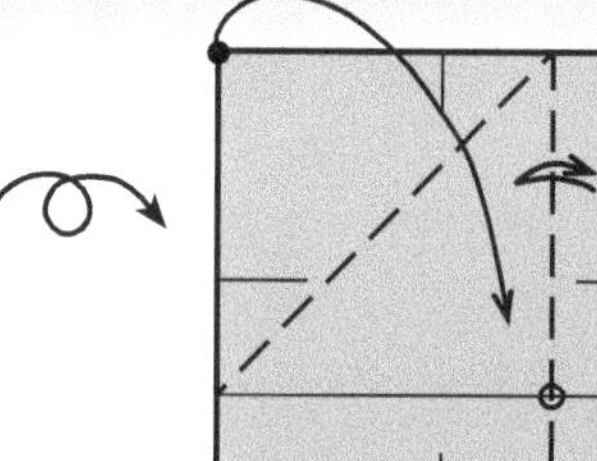

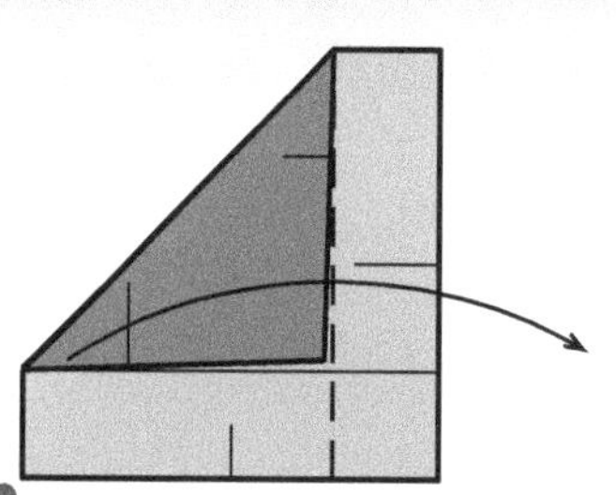

1 Pinch midpoints of each edge. Fold the bottom one quarter upwards and unfold.

2 Crease the right quarter. Fold the top left corner to the intersection of the horizontal and vertical creases.

3 Fold the flap to the right using the existing crease.

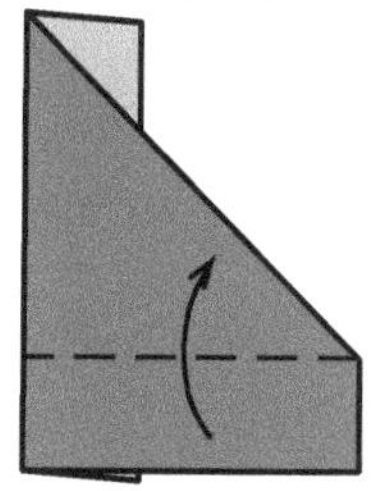

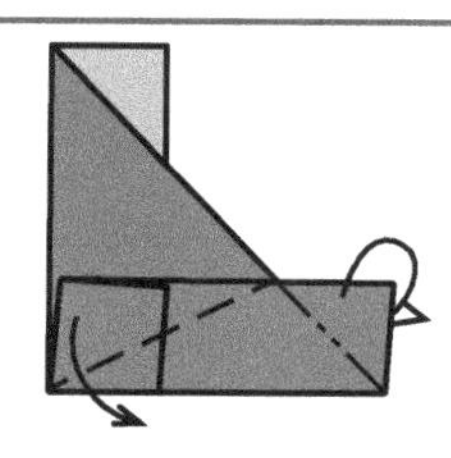

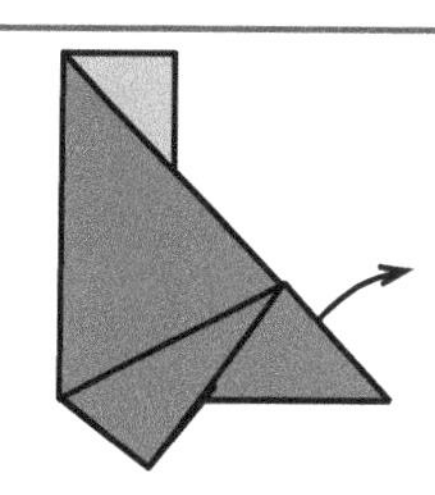

4 Valley fold the bottom flap up using the existing crease.

5 Fold the right corner behind. Fold the left corner down on the diagonal of a 1:2 rectangle.

6 Unfold the right corner from behind.

Assembly ★★

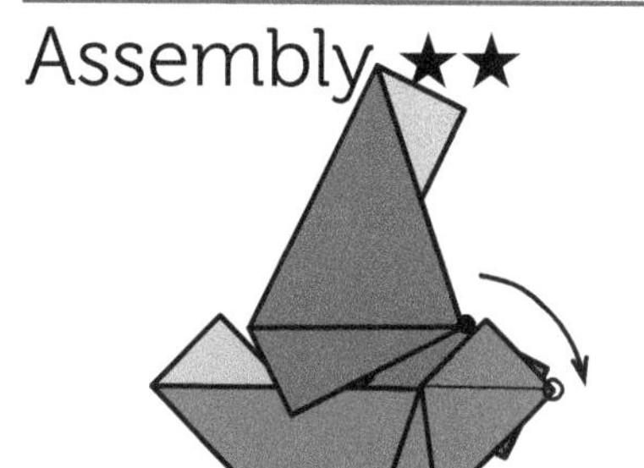

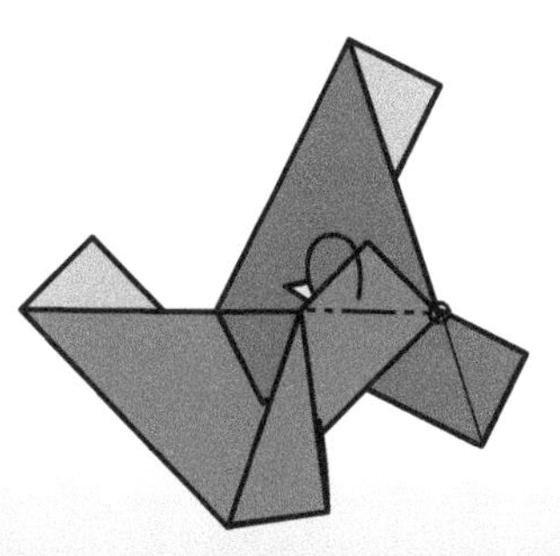

1 Hook the folded flap of the second unit between the first unit's layers and slide right.

2 Mountain fold the flap of the first unit behind to finish joining the first two units.

3 The first two units joined.

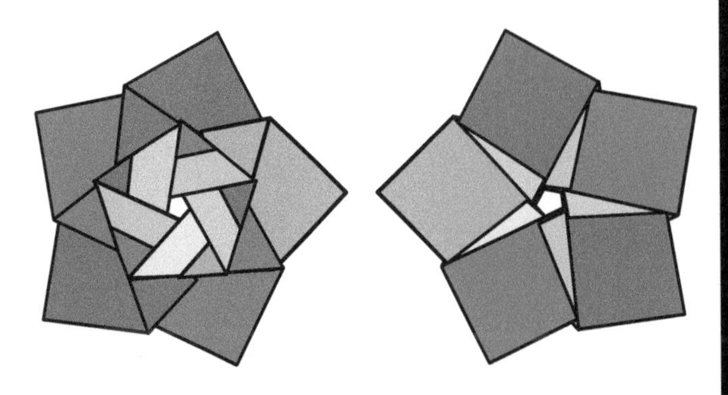

Star 5

5 by 6 Rectangle From Square

★★

Use five squares about 100 mm
Finished model ratio 1.6

A 4 by 4 square contains a 3:4 rectangle. The excess paper could be folded away but is left alone in this version. This model makes a good pinwheel when a pencil is put through the hole.

Module ★★

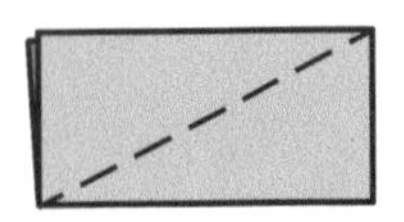

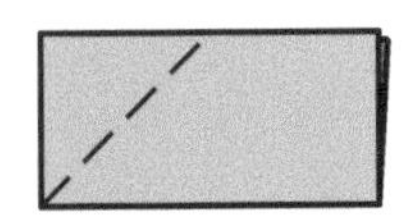

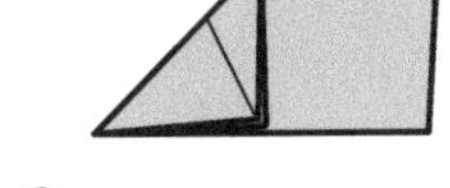

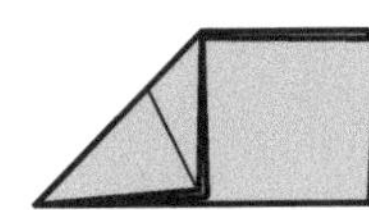

1 Start with a square book folded in half. Fold the diagonal of the 1:2 rectangle, creasing one layer only. Unfold and turn over.

2 Bisect the lower left right angle, folding through both layers.

3 Unfold and turn over.

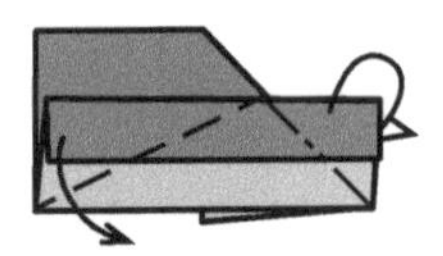

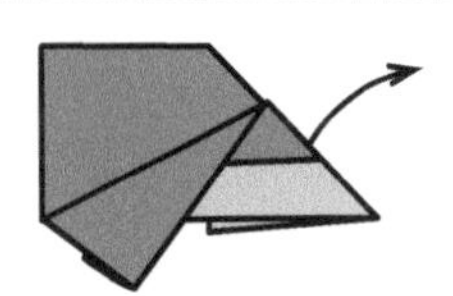

4 Valley fold the upper layer down one third. The fold goes through the intersection of creases.

5 Fold the right corner behind. Fold the left corner down using the diagonal of a 1:2 rectangle.

6 Unfold the right corner from behind.

Assembly ★★

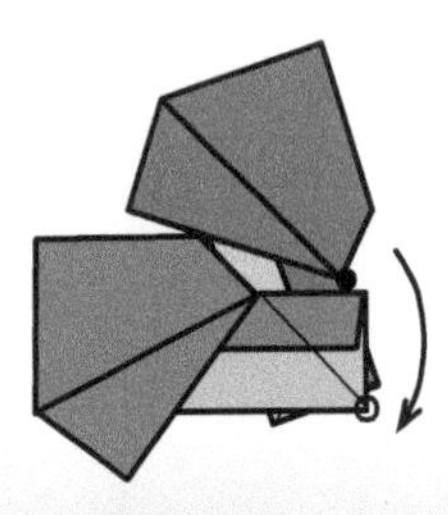

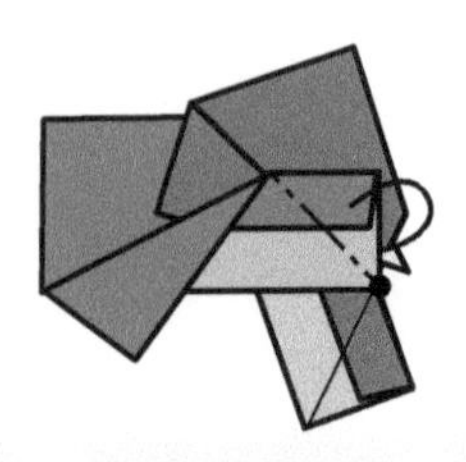

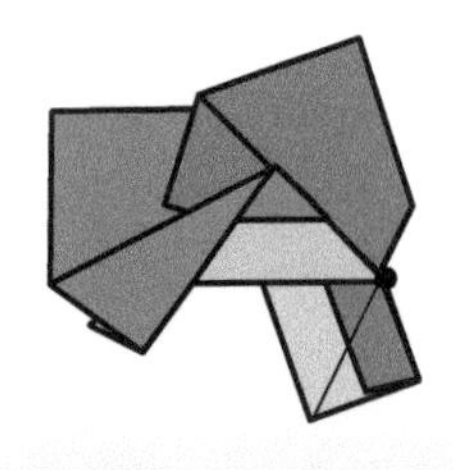

1 Hook the folded flap of the second unit between the first unit's layers and slide right.

2 Mountain fold the flap of the first unit behind to finish joining the first two units.

3 Mountain fold the flap of the first unit behind to finish joining the first two units.

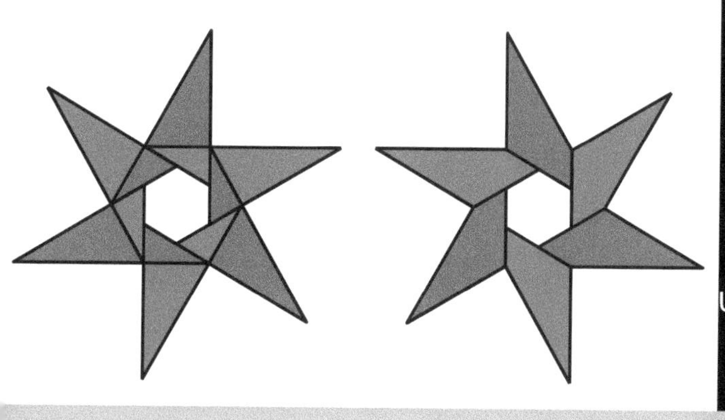

Poly Diag Star 6

★★

Use six quarter bronze rectangles, e.g. cut a bronze rectangle from A4 and then cut into quarters

Finished model ratio 3.5

The layers become a little thick when assembled so use larger or thinner paper than usual. The hook and lock assembly method is familiar from previous models.

Module ★★

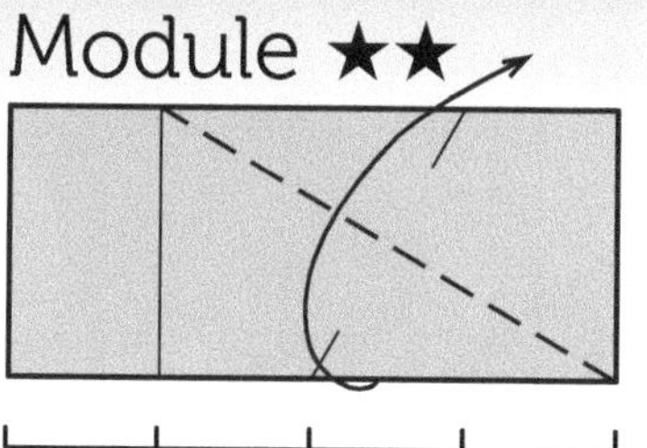

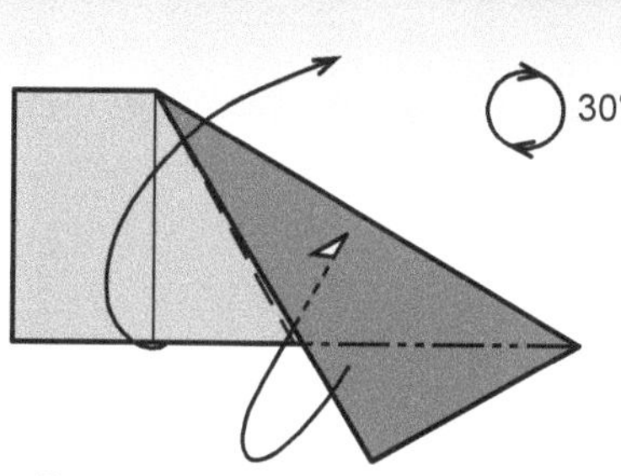

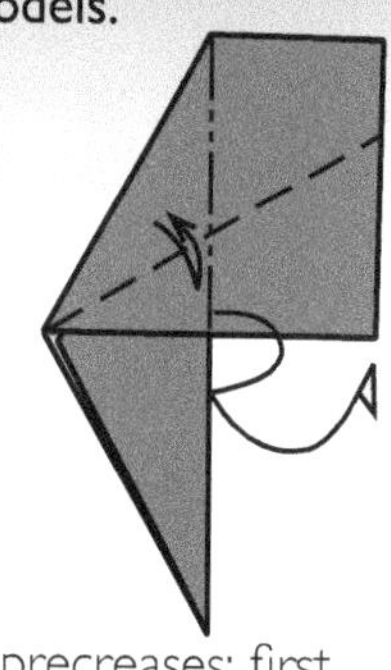

1 Crease one quarter from the left. Fold the diagonal of the larger bronze rectangle.

2 Fold on the raw edges. Tuck the lower flap inside. Rotate 30° clockwise.

3 Two precreases: first, mountain crease on the folded edge. Then bisect 60°.

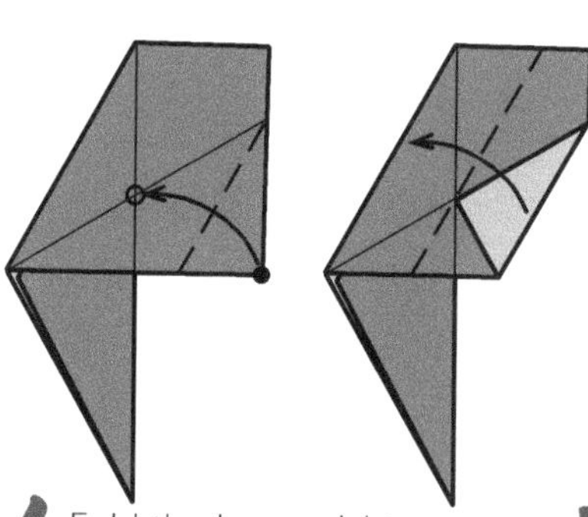

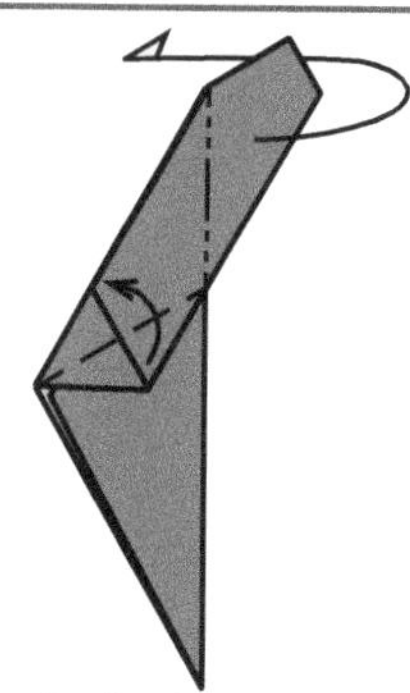

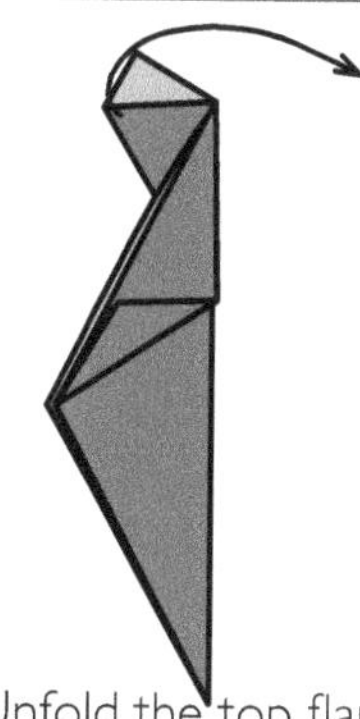

4 Fold the lower right corner to the intersection of precreases. Then fold over again.

5 Mountain fold the top right flap behind. Fold the little flap upwards.

6 Unfold the top flap.

Assembly ★★

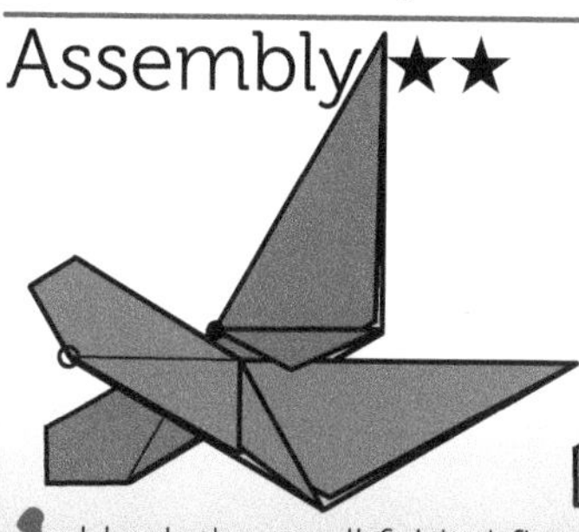

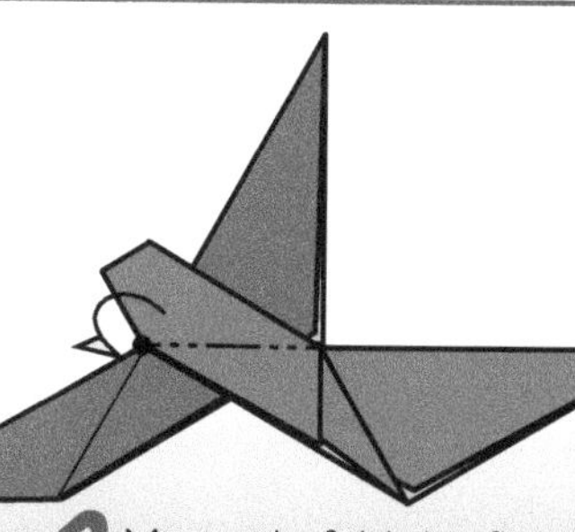

1 Hook the small folded flap of the second unit between the layers of the first unit.

2 Mountain fold the flap of the first unit behind to finish joining the first two units.

3 The first two units joined.

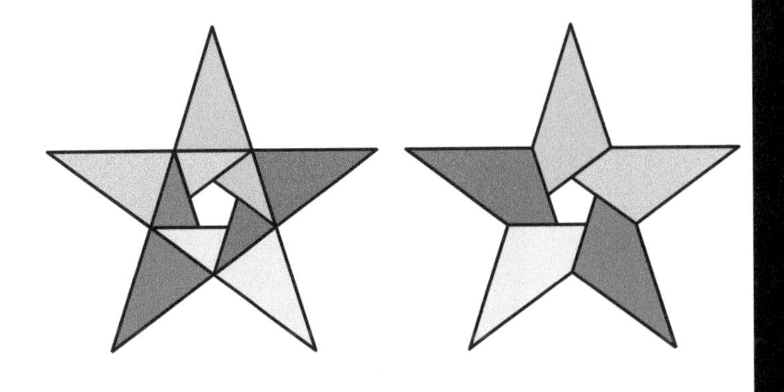

Poly Diag Star 5

★★

Use five 3:√2 rectangles, e.g. divide the long edge of A4 into thirds and cut

Finished model ratio 2.7

This is a five-piece variation of the six-piece version. The overall folding sequence is similar, although some location points differ.

Module ★★

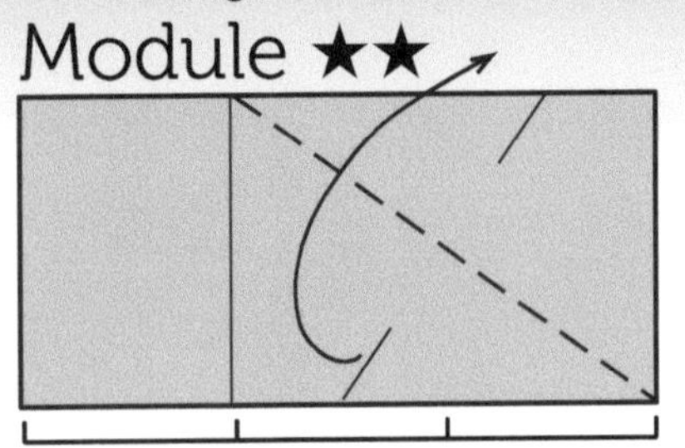

1 Crease one third from the left. Fold the diagonal of the larger silver rectangle.

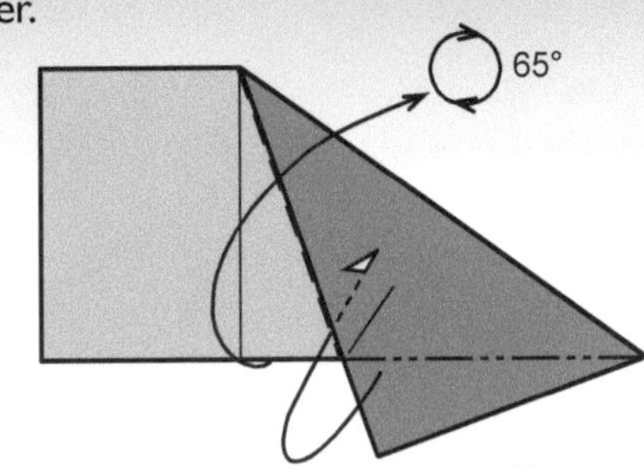

2 Fold on the raw edges. Tuck the lower flap inside. Rotate 65° clockwise.

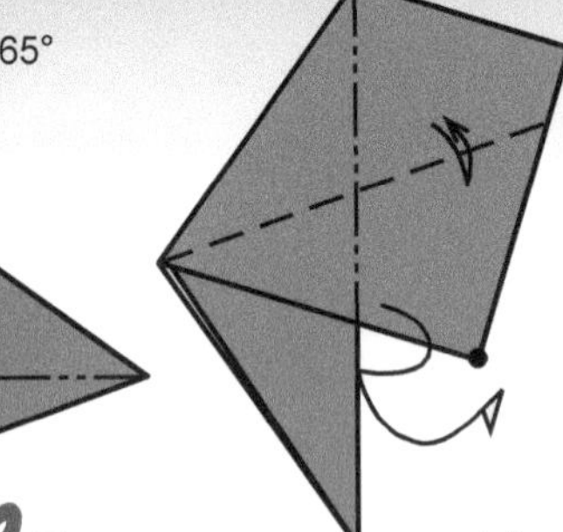

3 Two precreases: mountain crease on the folded edge. Then bisect lower left angle.

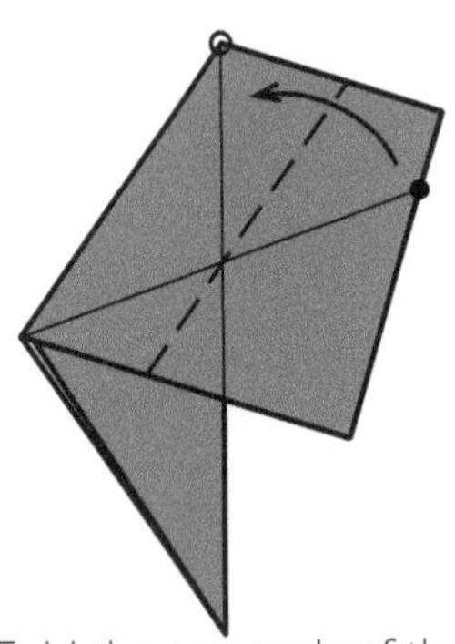

4 Fold the top ends of the creases together. The fold goes through the intersection of precreases.

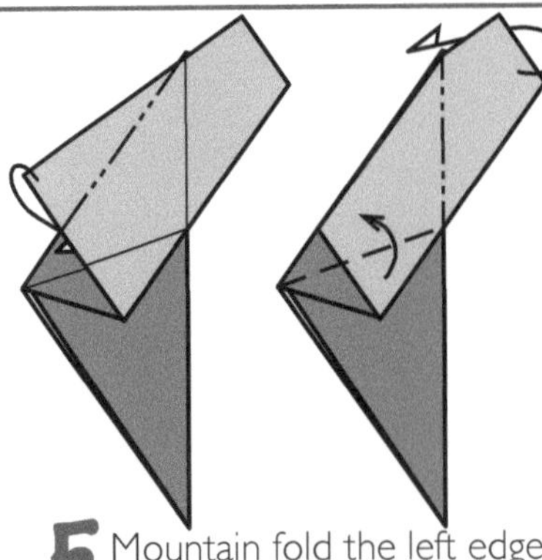

5 Mountain fold the left edge inside and the top right flap behind. Fold the little flap upwards.

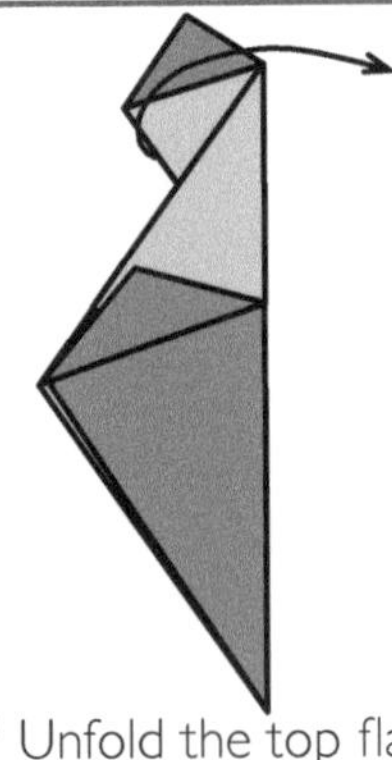

6 Unfold the top flap.

Assembly ★★

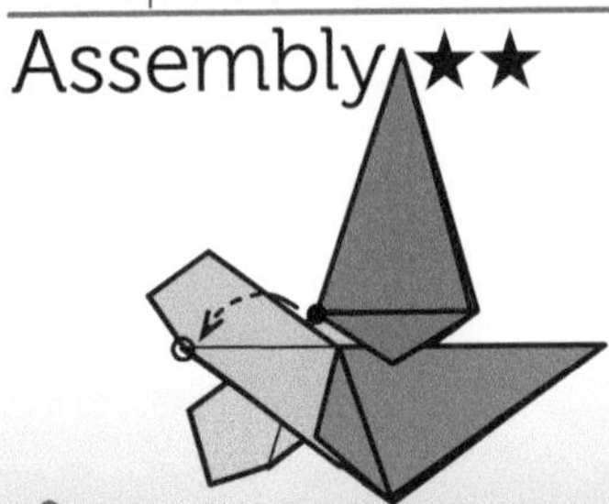

1 Hook the small folded flap of the second unit between the layers of the first unit.

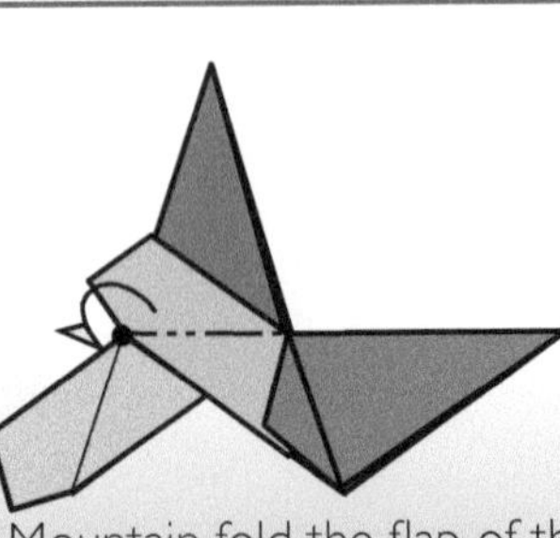

2 Mountain fold the flap of the first unit behind to finish joining the first two units.

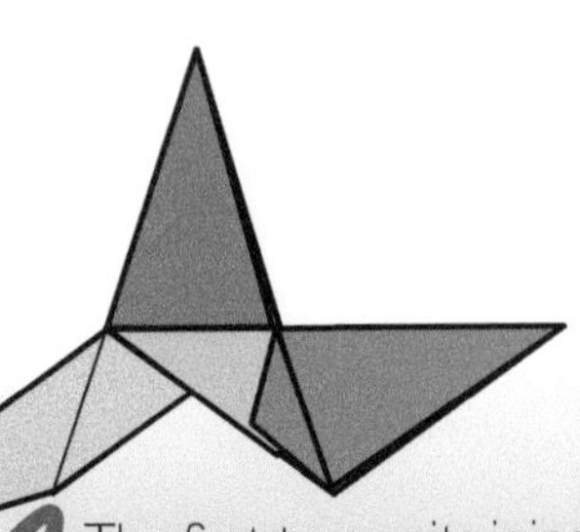

3 The first two units joined.

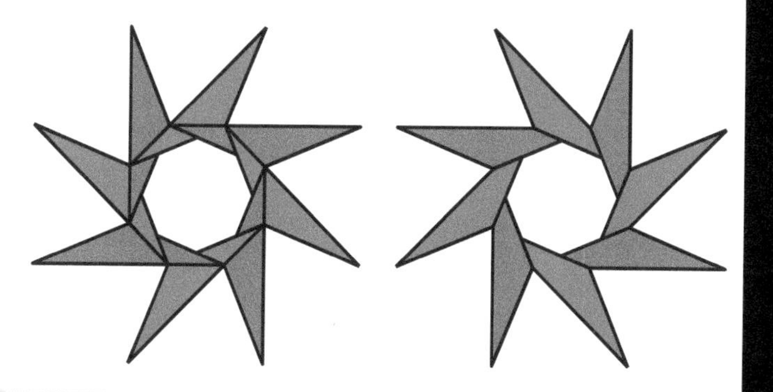

Poly Diag Star 8

★★

Use eight 1:2√2 rectangles,
e.g. divide the short edge of A4
into two and cut
Finished model ratio 3.5

This is an eight-piece variation of the six- and five-piece versions. The folding sequence is a little different to take advantage of the 22.5° geometry.

Module ★★

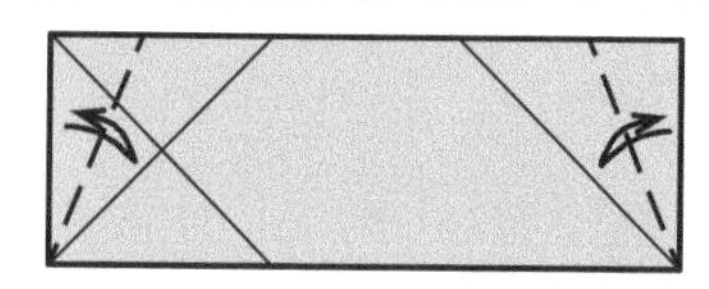

1 Bisect the right angles except the top right. Then bisect the left and right 45° angles.

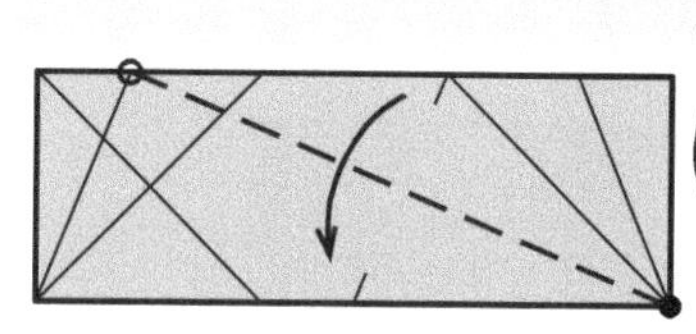

2 Fold the diagonal of the landscape leftover rectangle. You can make location marks by creasing between the circles.

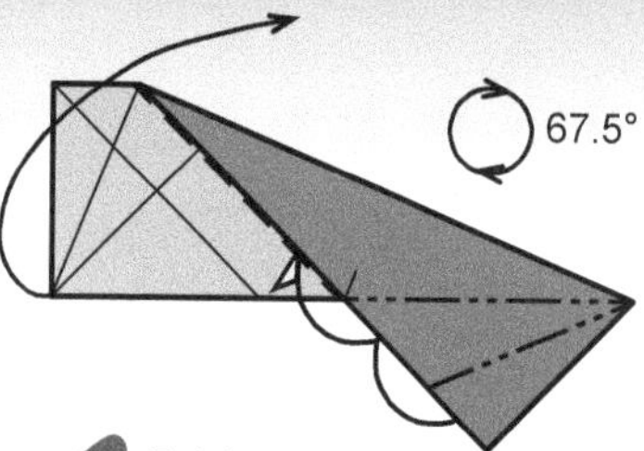

3 Fold on existing creases. Mountain fold the lower flap twice, tucking it inside. Rotate 65° clockwise.

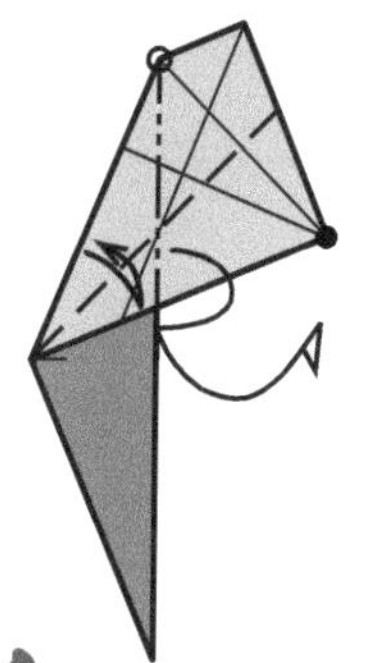

4 Two precreases: mountain crease on the folded edge. Then bisect 45°.

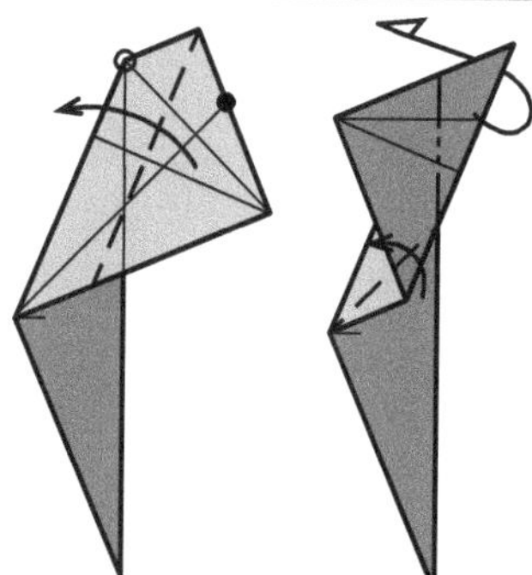

5 Valley fold on the 45° crease. Mountain fold the top right flap behind. Fold the little flap upwards.

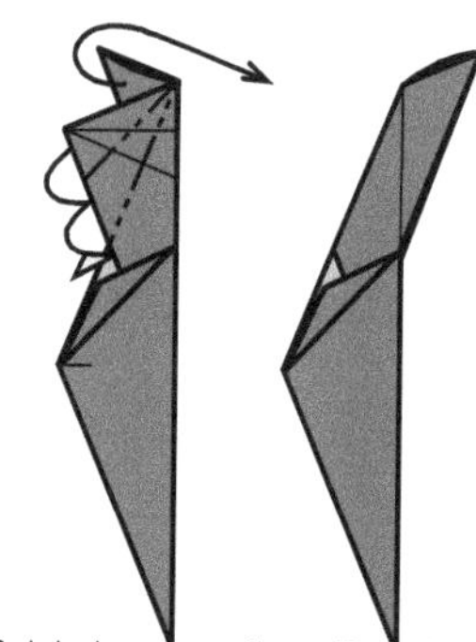

6 Unfold the top flap. Double fold the small right angled flap and tuck inside.

Assembly ★★

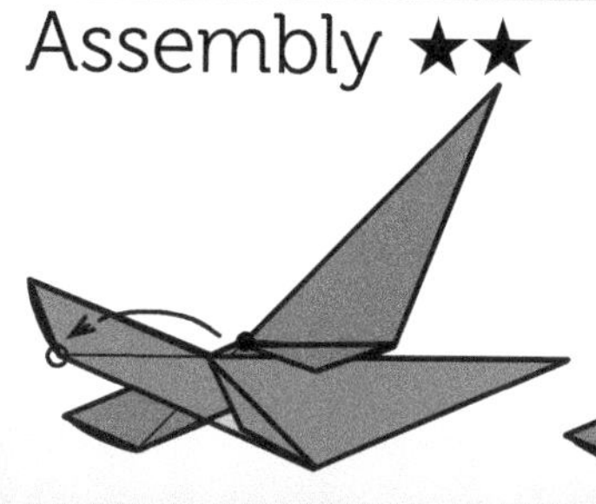

1 Hook the small folded flap of the second unit between the layers of the first unit.

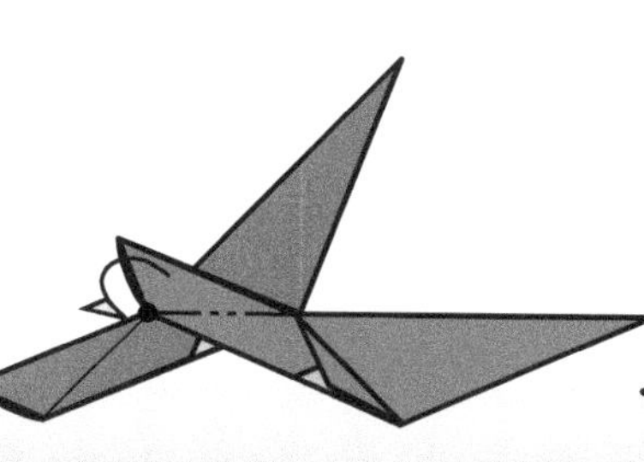

2 Mountain fold the flap of the first unit behind to finish joining the first two units.

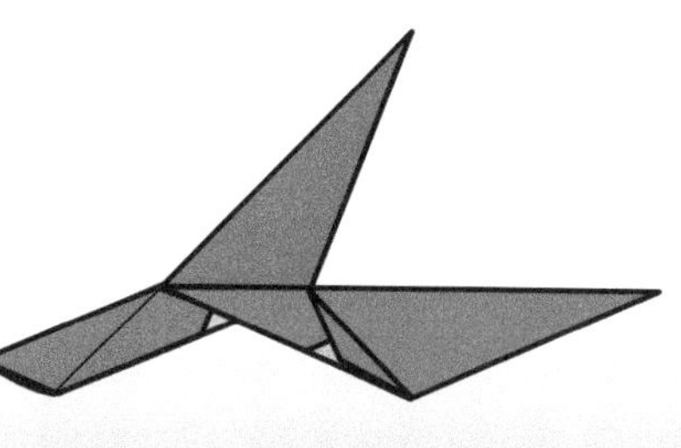

3 The first two units joined.

Boat Unit 5 Pentagram

★★

Use five A7 rectangles
Finished model ratio 1.3

Although the layers become thick in the finished model, the result is sturdy and can be thrown with a flick of the wrist. The strong assembly can be a surprise given the small size of the locking tips in step 6: the strength comes from the multiple layers in module step 5.

Module ★★

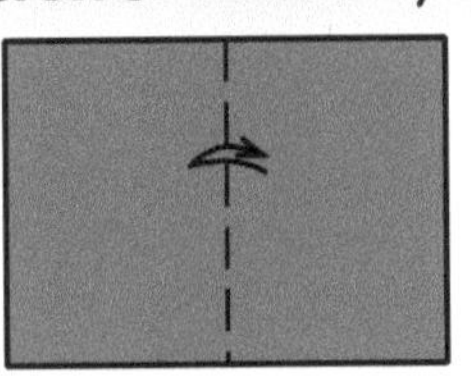

1 Fold in half bringing short edges together. Unfold and turn over.

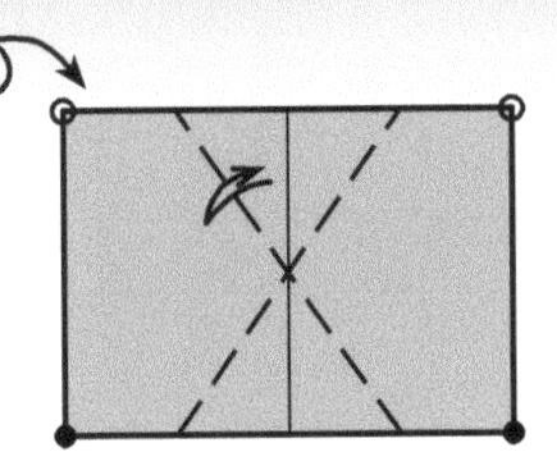

2 Fold opposite corners together and unfold.

3 Use the existing creases to fold in half, making two inside reverse folds, like a waterbomb base. Rotate 90° anticlockwise.

4 Fold left then right flaps.

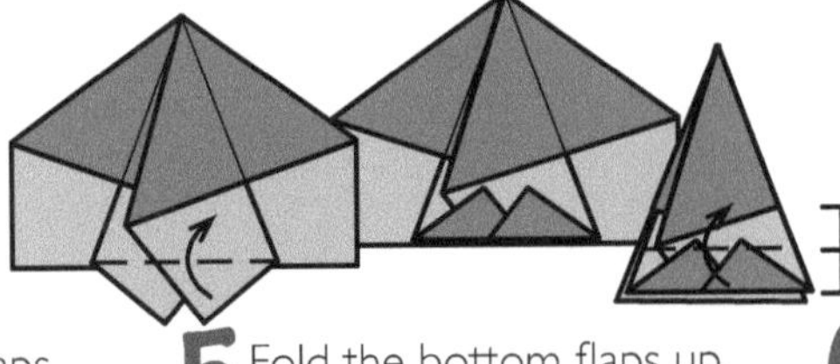

5 Fold the bottom flaps up. Repeat behind. Fold the bottom flaps up again.

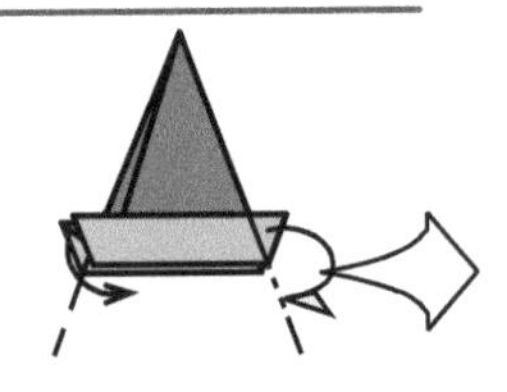

6 Mountain fold the right tip against the folded edge behind and unfold. Turn over and repeat.

Assembly ★★

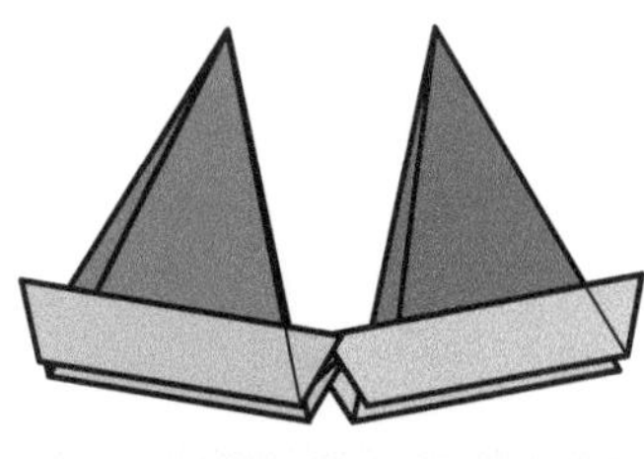

1 Interleave the flaps of both units. The second unit is at the bottom, and the first unit is at the top.

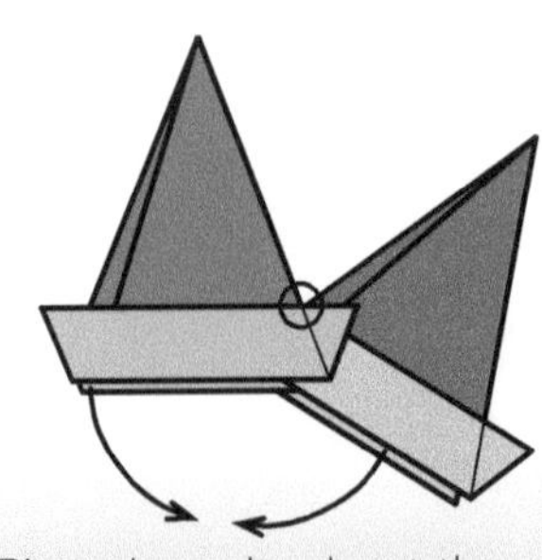

2 Pivot the units about the circled point.

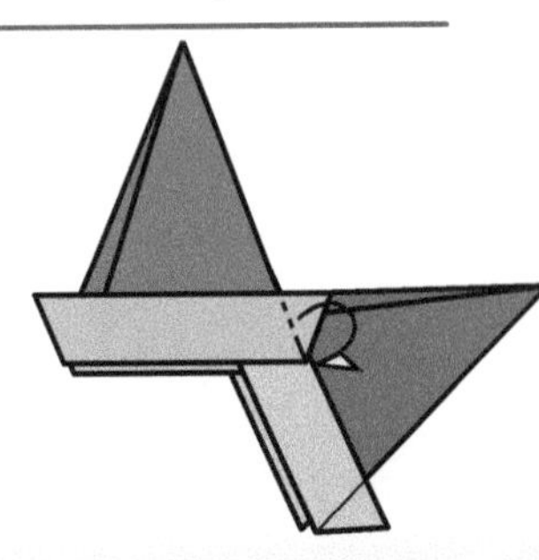

3 Tuck the right tip of the first unit into the second unit. Turn over and repeat.

Boat Unit 6 Hexagram

★★

Use six bronze rectangles with short edges about 70 mm

Finished model ratio 1.7

This unit is similar to WXYZ but folded from a bronze rectangle instead of a square. The unit looks like a paper boat, even though the opening is in the wrong place for a boat.

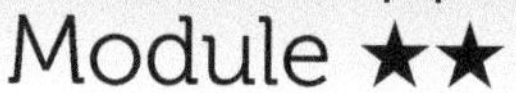

Module ★★

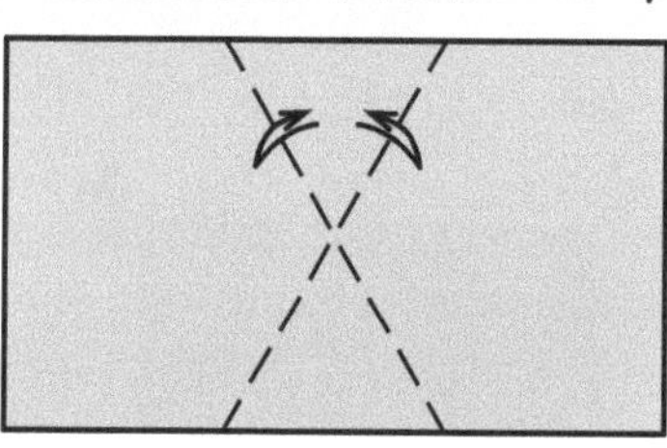

1 Fold and unfold each corner to the opposite corner. Fold in half bringing short edges together and unfold. Turn over.

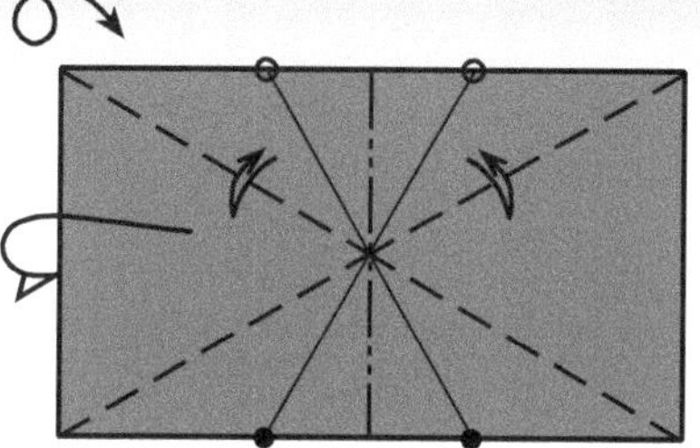

2 Fold and unfold both diagonals using the ends of the creases from step 1. Then mountain fold in half. Rotate 90° clockwise.

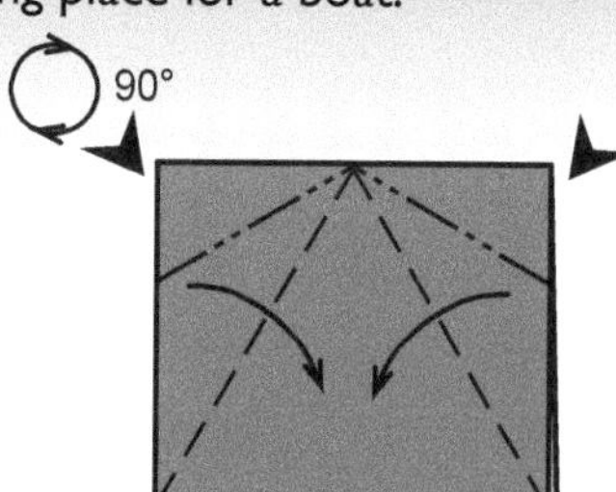

3 Squash fold the top left and right corners.

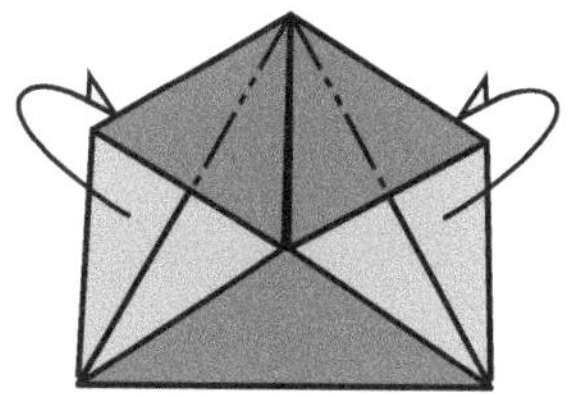

4 Mountain fold the left and right flaps behind.

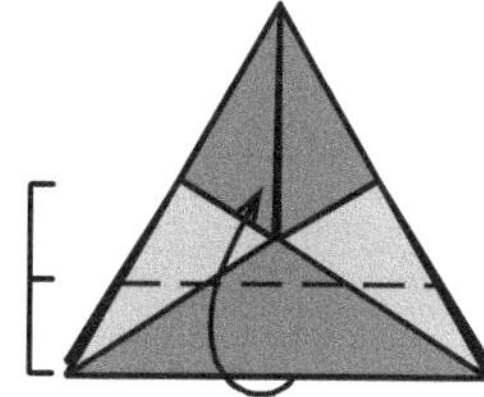

5 Fold the bottom edge upwards to meet the corner of the squashed flap. Repeat behind.

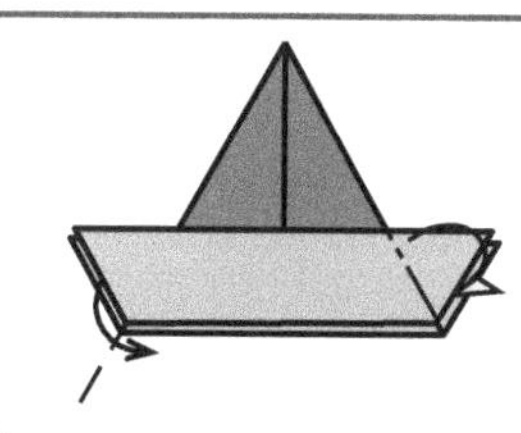

6 Mountain fold the right tip against the folded edge behind and unfold. Turn over and repeat.

Assembly ★★

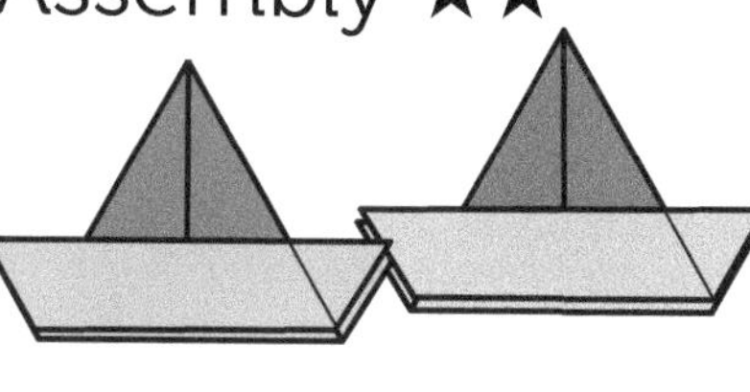

1 Interleave the flaps of both units. The second unit is at the bottom, and the first unit is at the top.

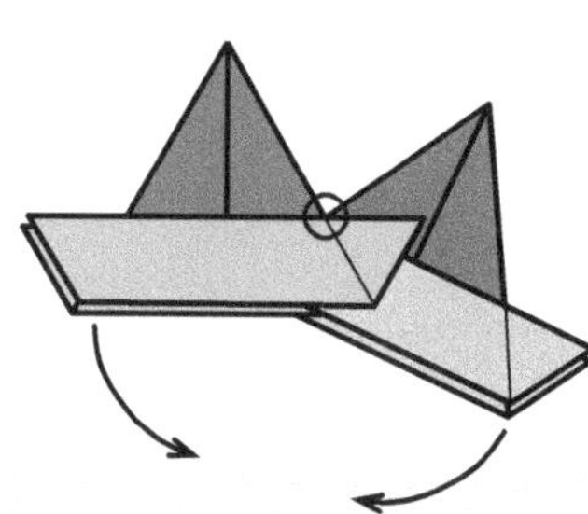

2 Pivot the units about the circled point.

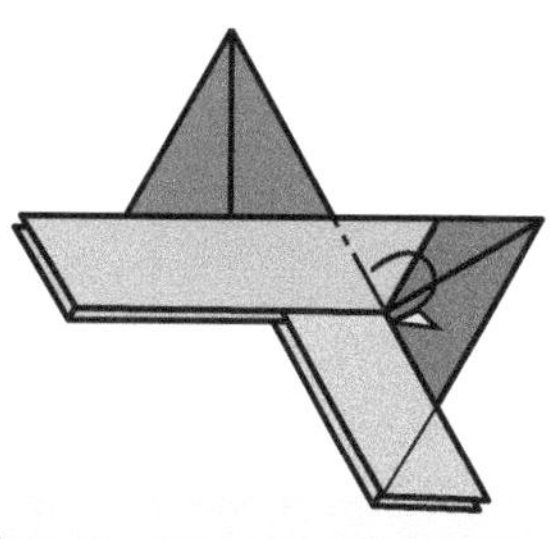

3 Tuck the right tip of the first unit into the second unit. Turn over and repeat.

Boat Unit 7 Heptagram

★★
Use seven 4:5 rectangles
with short edges about 100 mm
Finished model ratio 2.0

A prime number of units can look good in a single colour for all units. To make a 4:5 rectangle from a square, cut off one fifth. From A4, cut off a 35 mm strip, then cut into quarters. You could also divide the short edge into quarters and then find five quarters on the long side.

Module ★★

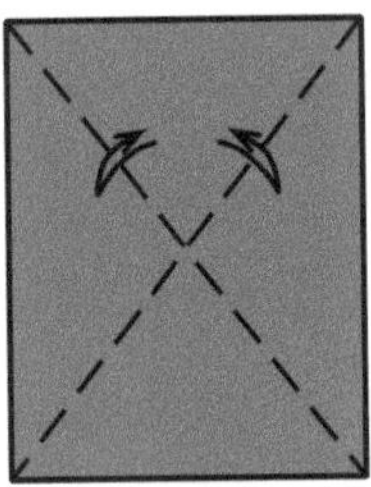

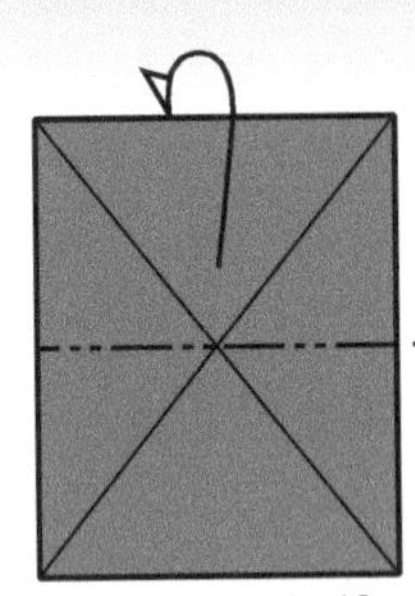

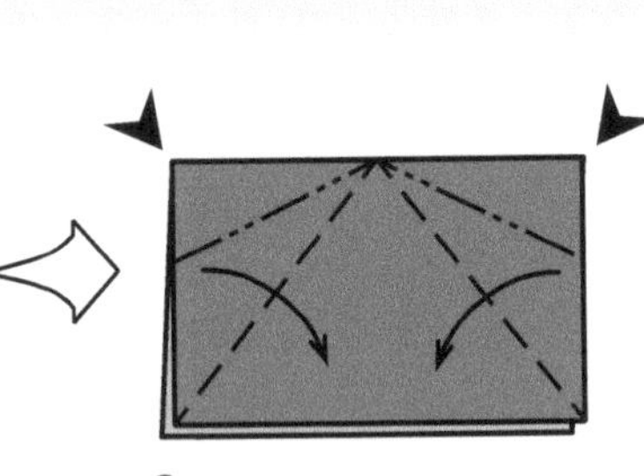

1 Fold and unfold each corner to the opposite corner.

2 Mountain fold in half.

3 Squash fold the top left and right corners.

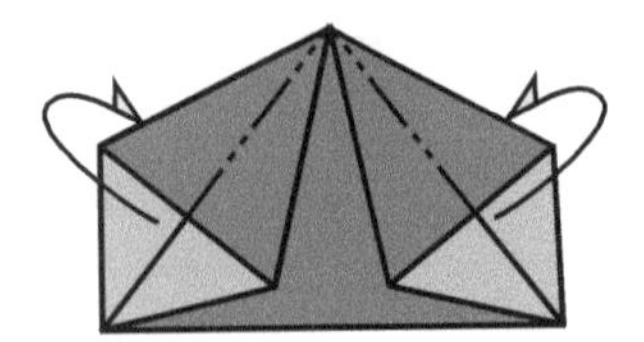

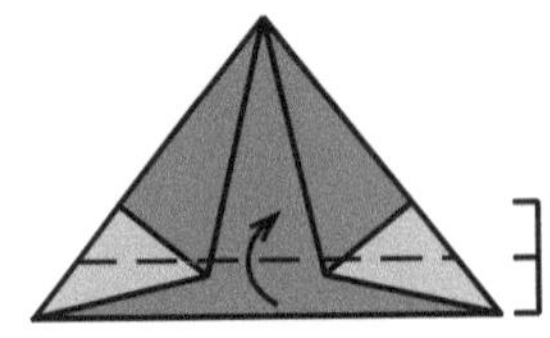

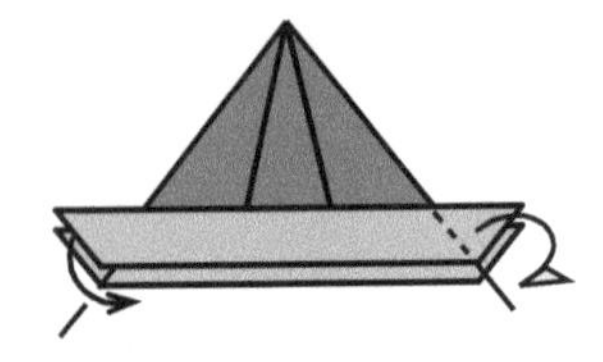

4 Mountain fold the left and right flaps behind.

5 Fold the bottom edge upwards to meet the corner of the squashed flap. Repeat behind.

6 Mountain fold the right tip against the folded edge behind and unfold. Turn over and repeat.

Assembly ★★

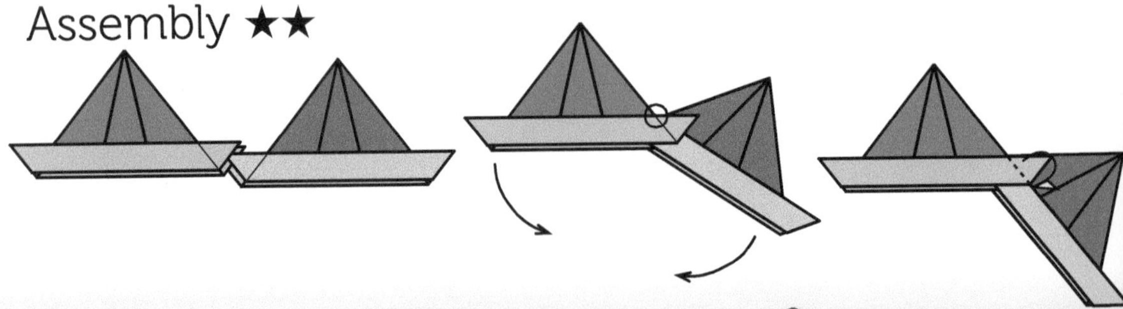

1 Interleave the flaps of both units. The second unit is at the bottom, and the first unit is at the top.

2 Pivot the units about the circled point.

3 Tuck the right tip of the first unit into the second unit. Turn over and repeat.

Chapter 8

Wreaths

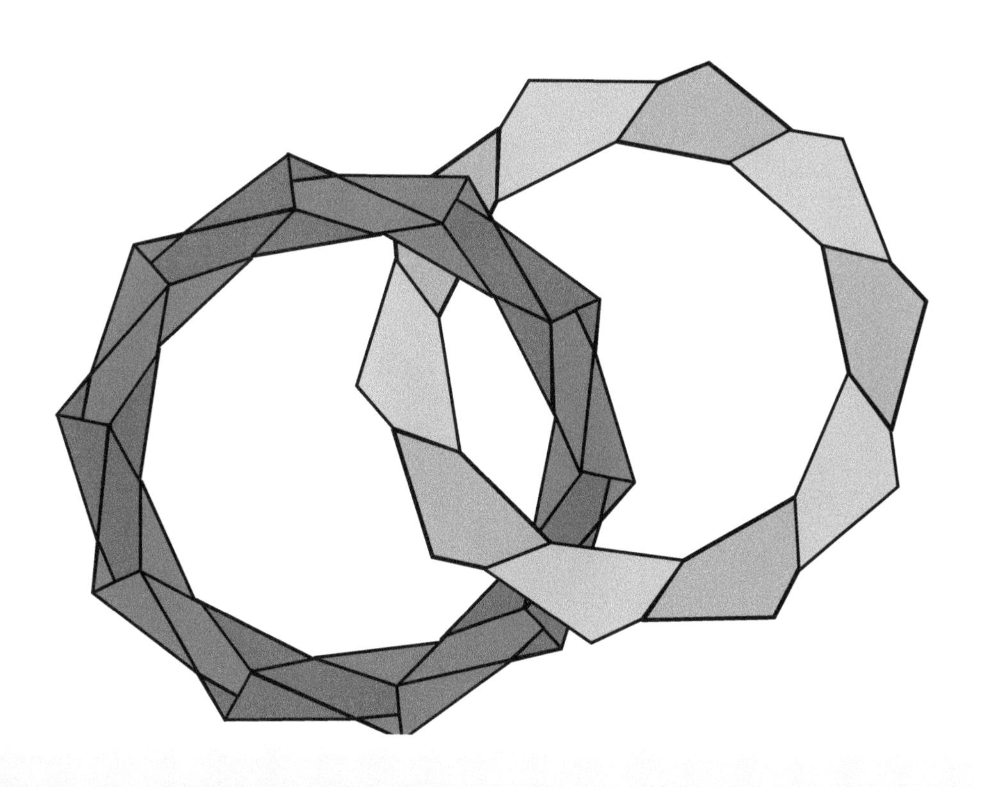

Swirl 10

Square From Silver Rectangle

★

Use ten silver rectangles
A7 or larger
Finished model ratio 2.5

The units only have folds from the procedure to make a square from rectangle. A single fold is added to each unit during assembly. It may be surprising at first that the two sides are so different. However, both sides only have rotational symmetry.

Module ★

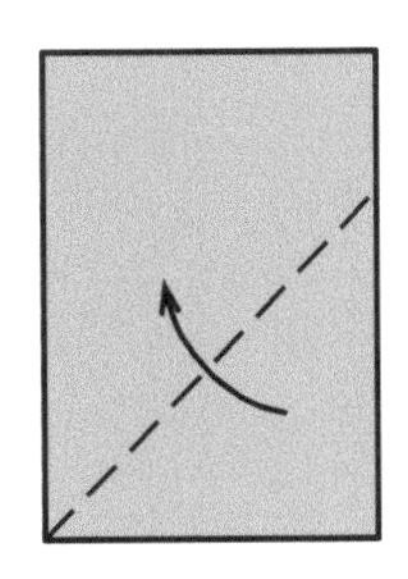

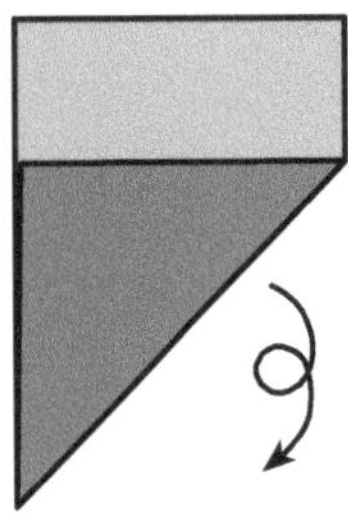

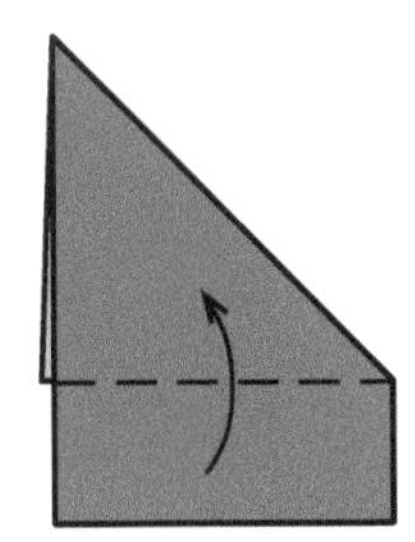

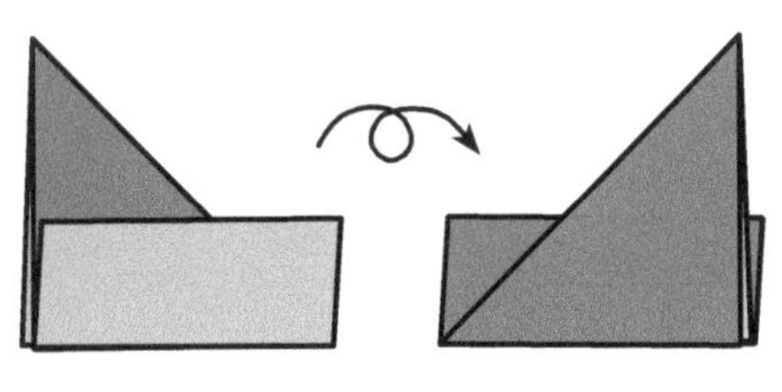

1 Bisect the lower left right angle. Turn over.

2 Fold the lower flap up so that the fold aligns with the raw edge behind it.

3 Turn over.

Assembly ★

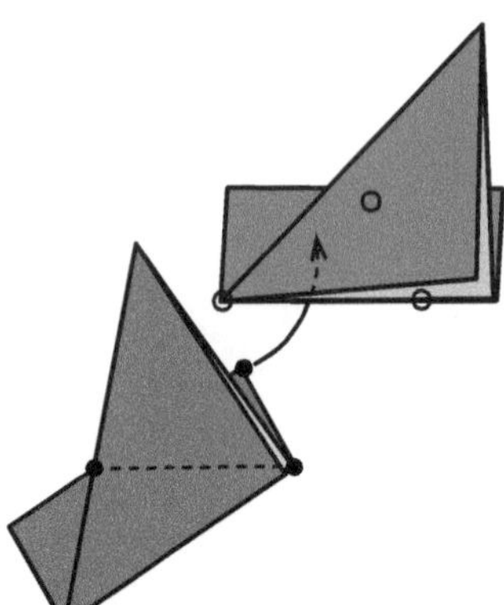

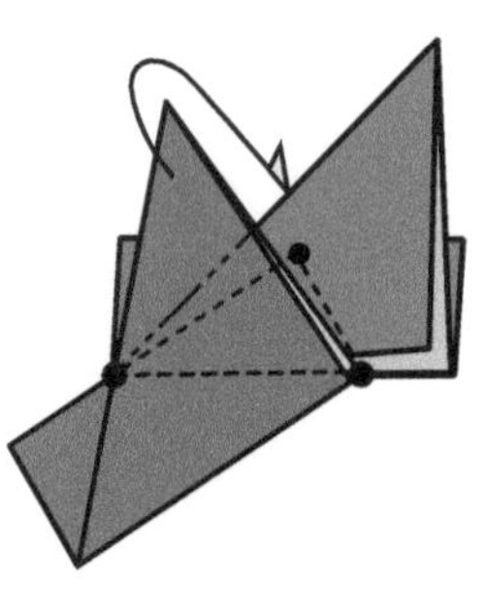

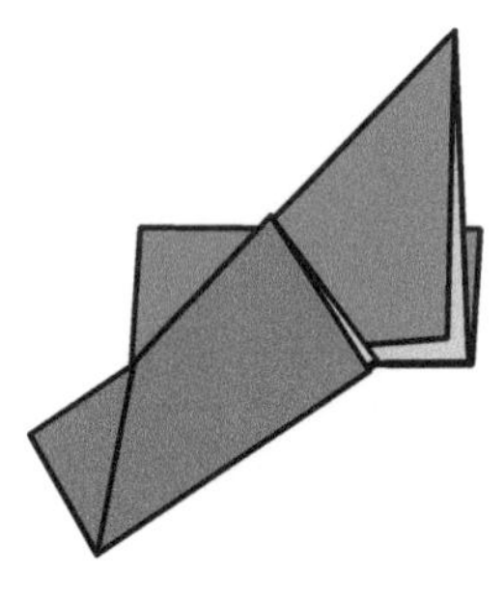

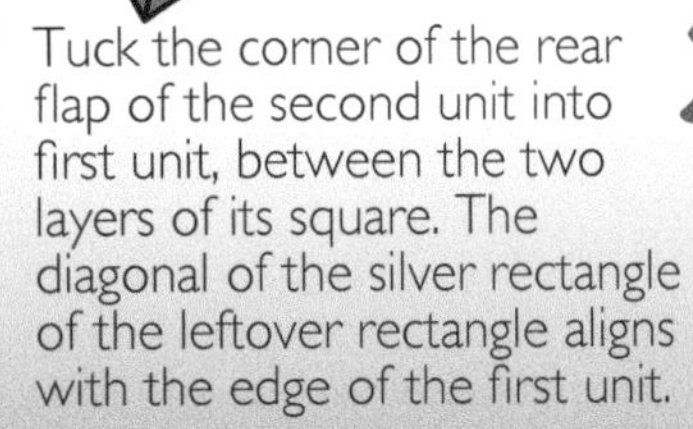

1 Tuck the corner of the rear flap of the second unit into first unit, between the two layers of its square. The diagonal of the silver rectangle of the leftover rectangle aligns with the edge of the first unit.

2 Mountain fold the flap of the second unit behind to lock.

3 Continue adding units in a clockwise direction.

4 Mountain fold the flap of the first unit behind to finish joining the first two units.

5 Two units joined.

6 Hook the third unit onto the second unit. Slide it down as far as it will go.

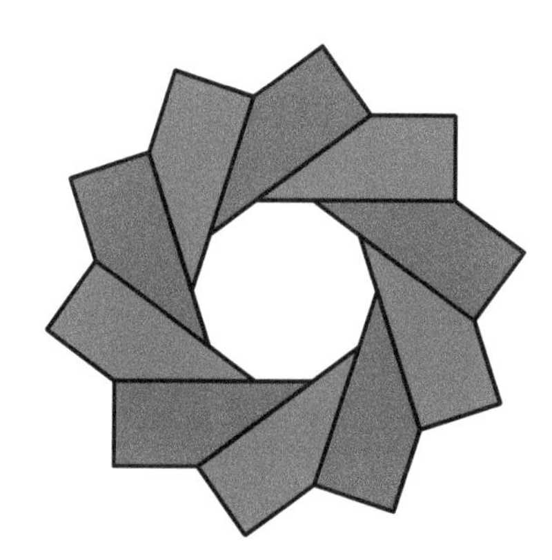

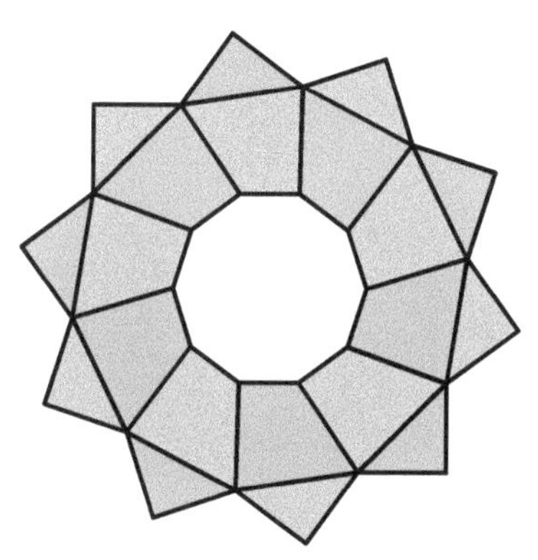

7 Mountain fold the flap to finish joining the third unit to the second.

8 Hook the fourth unit onto the third unit. Slide and mountain fold to lock. Unfold the small triangular flap of the first unit.

9 To join the fourth to the first, lift the end of the fourth unit up and over the first unit.

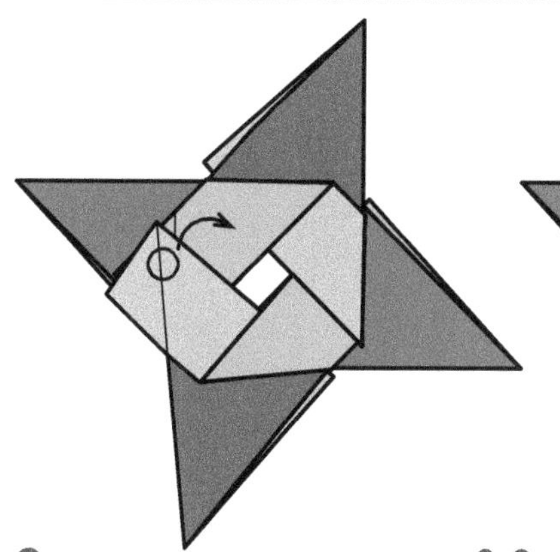

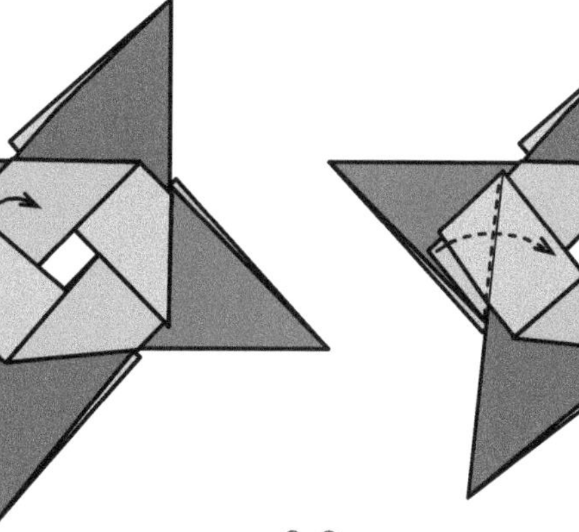

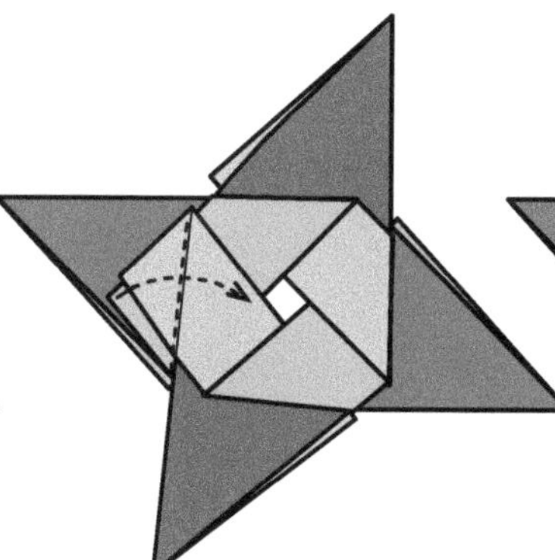

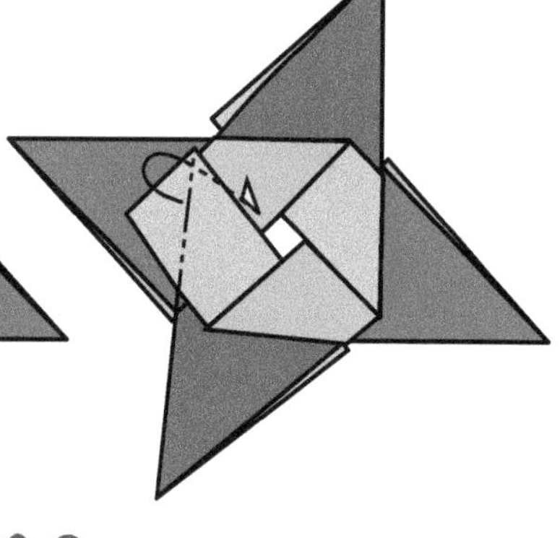

10 Hold the end of the fourth unit and move it to the right.

11 Valley fold the small flap of the first unit into the fourth unit.

12 Mountain fold the small flap of the fourth unit into the first unit to finish joining all units.

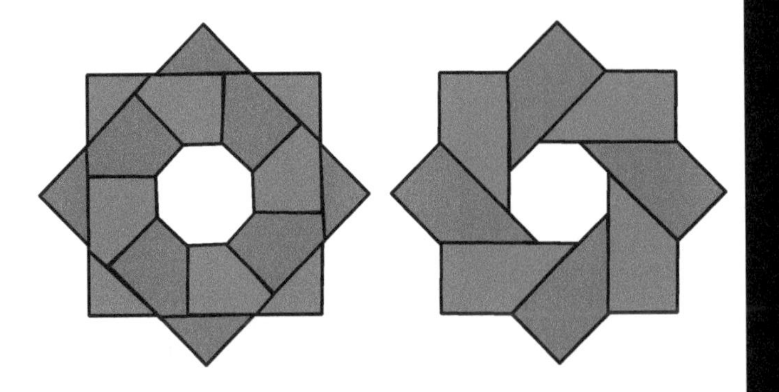

Octagram Wreath

★

Use eight 2:1 rectangles with long edges at least 75 mm
Finished model ratio 0.71

This eight-piece version of the previous model is even sturdier thanks to the 45° geometry. It could be made from 3:2 rectangles, but it is convenient to use 2:1 rectangles instead. The extra paper means each unit displays a single colour. You can make one model from one square of 15 cm origami paper: cut it into eight 2:1 rectangles. In fact you can speed the folding by precreasing a 4 by 4 grid, cutting in half, precreasing the fold in step 2 then cutting into quarters.

Module ★

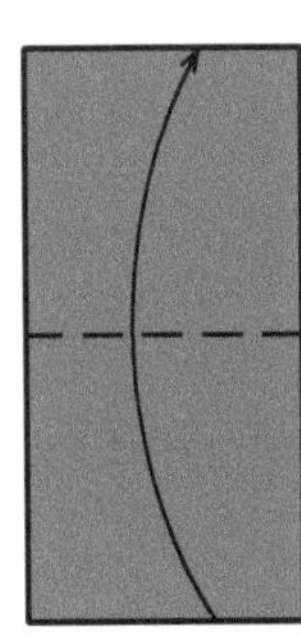

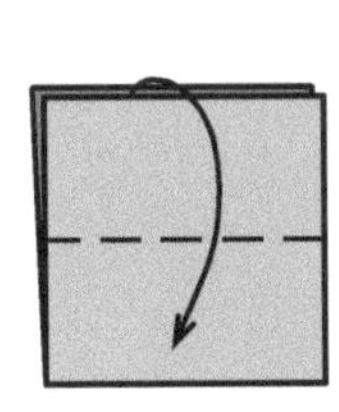

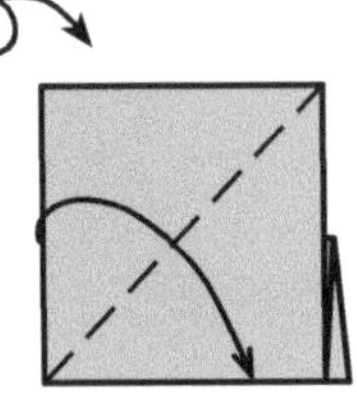

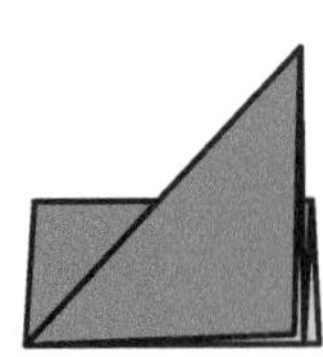

1 Fold in half, bringing the short edges together.

2 Fold the top layer in half. Turn over.

3 Fold the diagonal of the top layer square. Unit complete.

Assembly ★

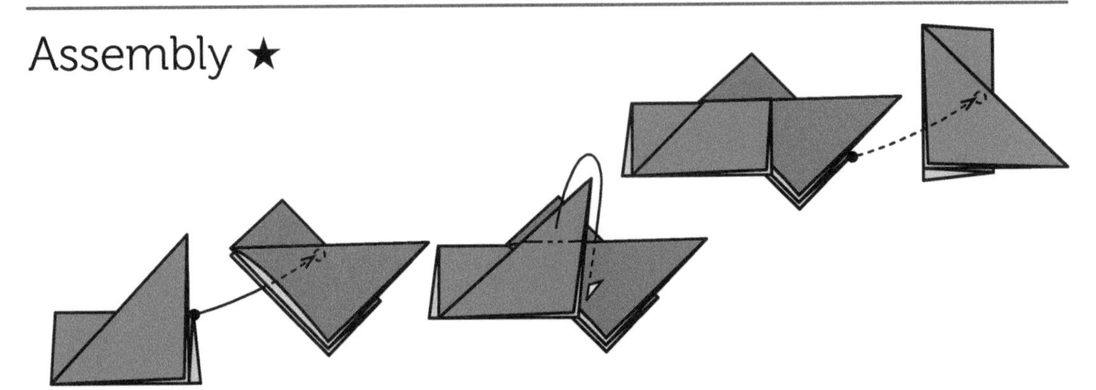

1 Tuck the top right corner of the first unit's rear flap between the top two layers of the second unit. Slide the first unit along the second unit's diagonal as far right as possible.

2 Fold the top flap behind and between the layers of the second unit.

3 Continue adding units in a clockwise direction.

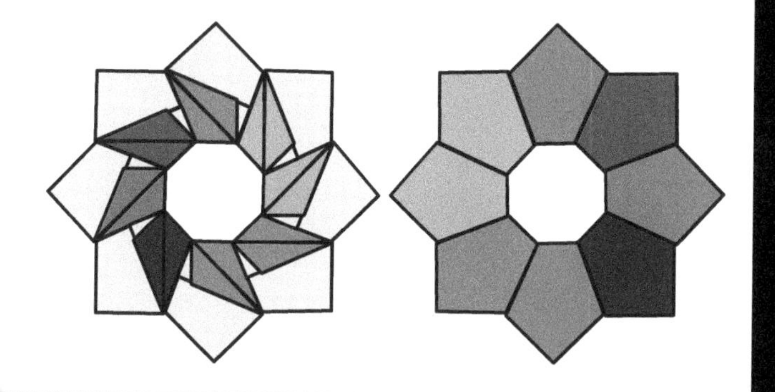

Octagram

Paper Cup
Kite Wheel

★★

Use eight 75 mm squares or larger
Finished model ratio 2.0

This model uses the geometry of the traditional paper cup. It was inspired by Dave Petty's explorations [Petty 94]. Try using a rainbow of colours.

Module ★★

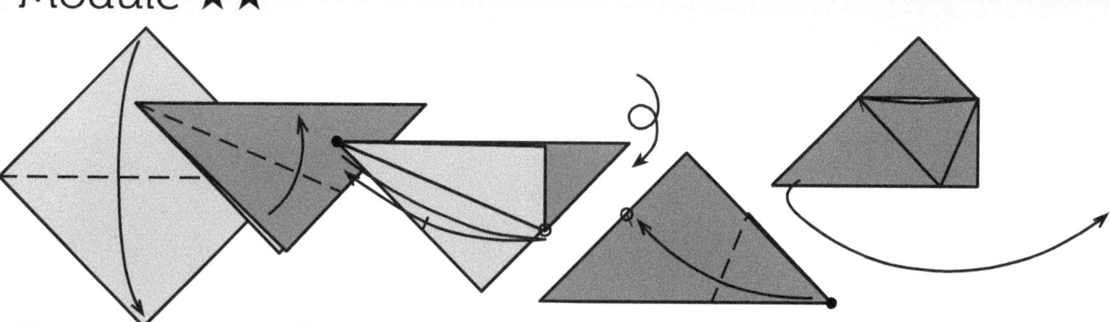

1 Crease the diagonal. Kite fold the top layer.

2 Pinch edge. Turn over.

3 Fold opposite corner to pinch mark.

4 Unfold left flap behind.

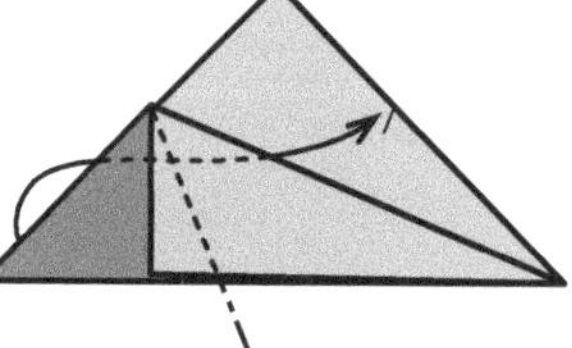

5 Inside reverse fold using the creases from step 3.

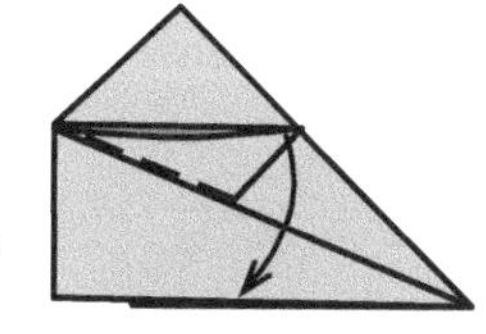

6 Fold down.

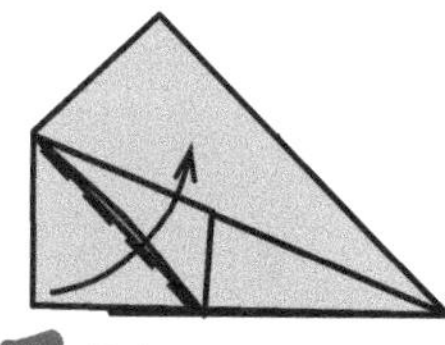

7 Fold up.

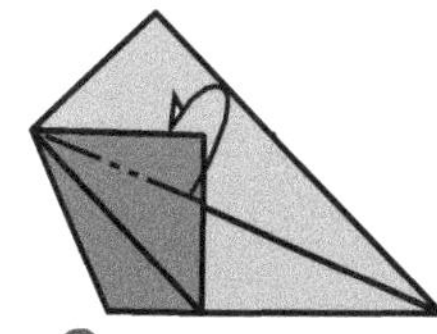

8 Fold behind and then unfold to step 6.

Assembly ★★

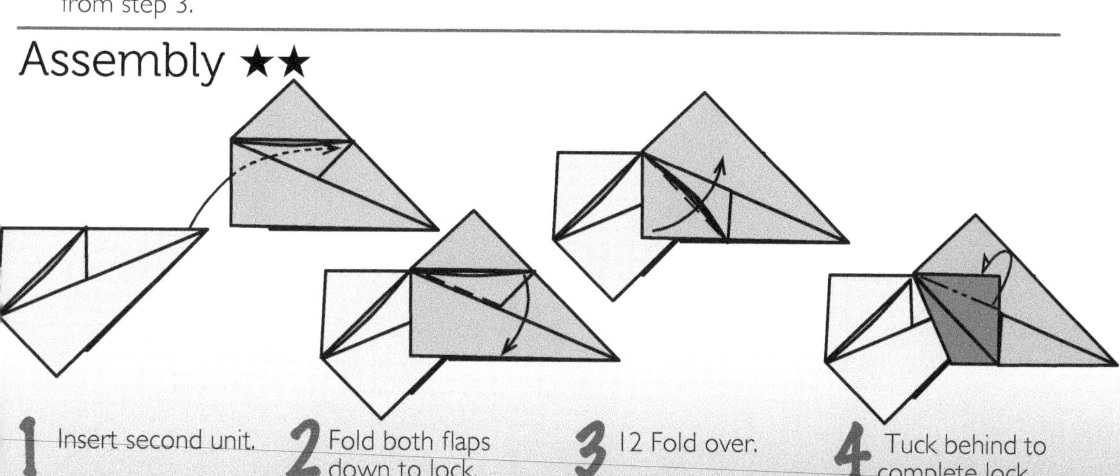

1 Insert second unit.

2 Fold both flaps down to lock.

3 12 Fold over.

4 Tuck behind to complete lock.

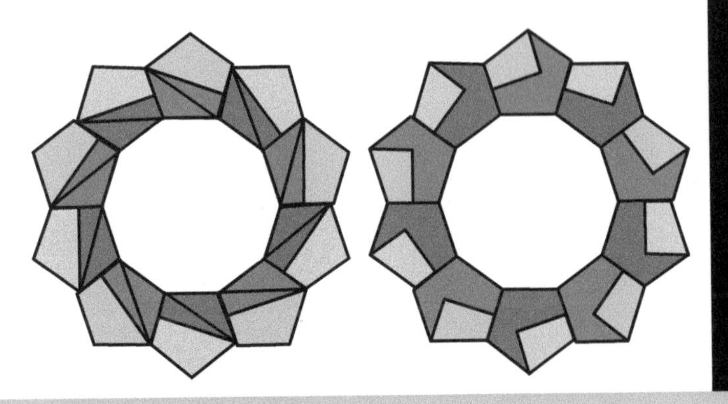

Kite Wheel 10

★★★
Use ten A7 rectangles or larger

Finished model ratio 2.8

This ten-piece variation of the previous model is assembled in a similar way. The layers can become thick, so use thin or large paper. Alternatively, cut off the flaps in step 2 to make the starting shape a rhombus.

Module ★★★

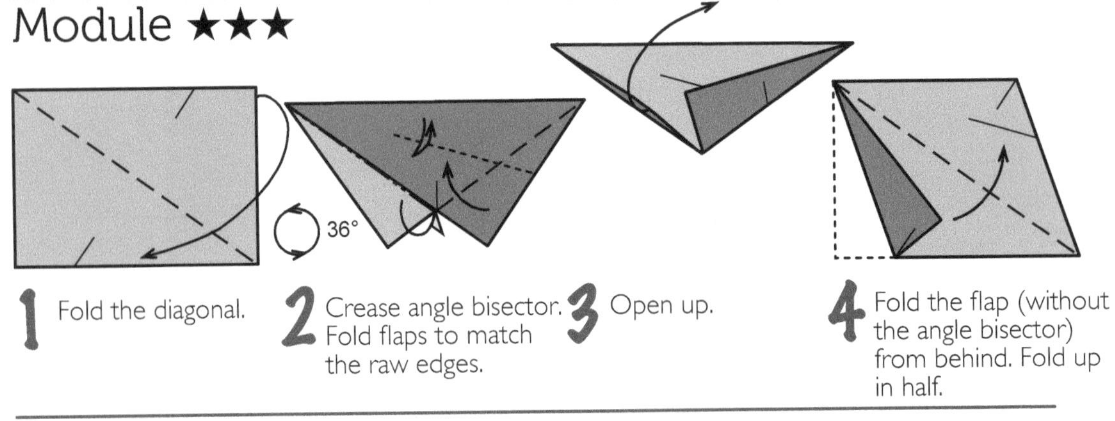

1 Fold the diagonal.

2 Crease angle bisector. Fold flaps to match the raw edges.

3 Open up.

4 Fold the flap (without the angle bisector) from behind. Fold up in half.

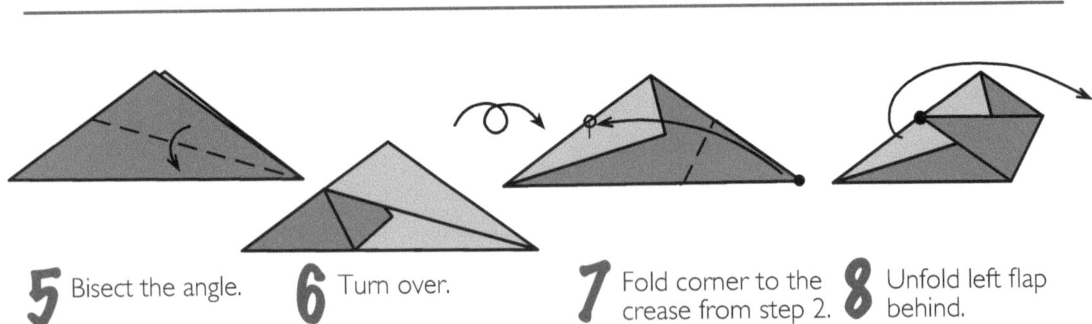

5 Bisect the angle.

6 Turn over.

7 Fold corner to the crease from step 2.

8 Unfold left flap behind.

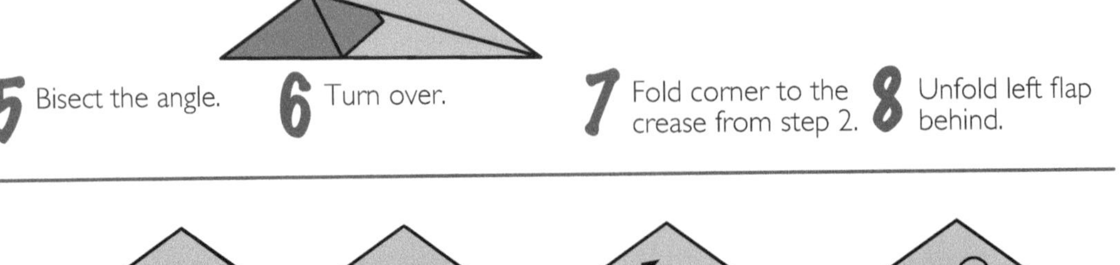

9 Inside reverse fold using the creases from step 3.

10 Fold down.

11 Fold up.

12 Fold behind and then unfold to step 10. To assemble, insert the second unit into the reverse fold. Do steps 10 to 12.

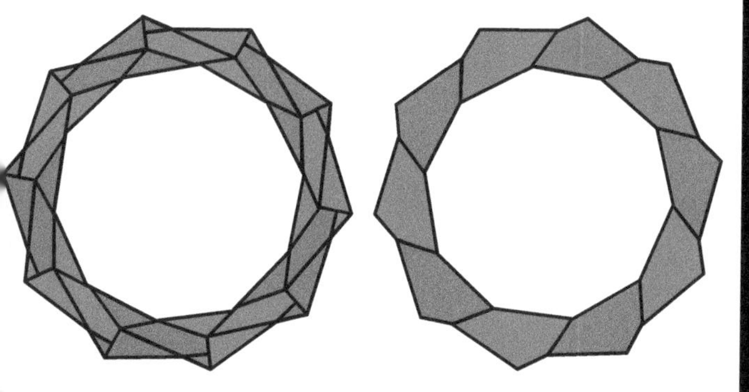

Decagon Ring

★

Use ten 75 mm squares or larger
Finished model ratio 2.8

This model's starting point is Nakano's Flying Saucer 4 [Nakano 85]. However, instead of using glue to join the units, a simple lock suffices. Try linking two rings or making a chain of rings.

Module ★

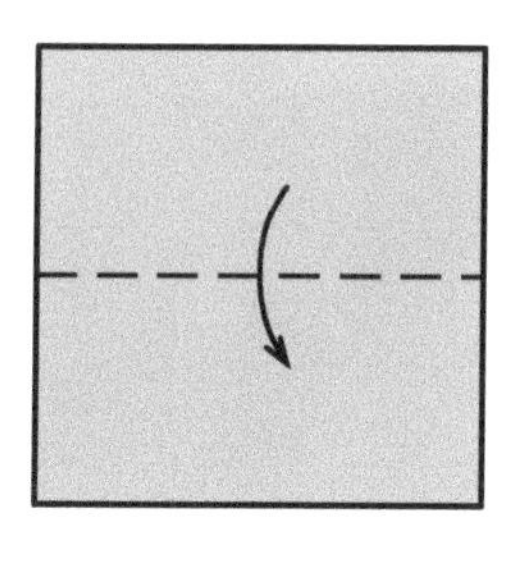

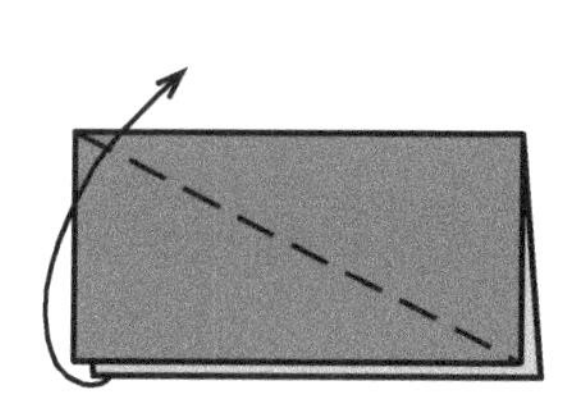

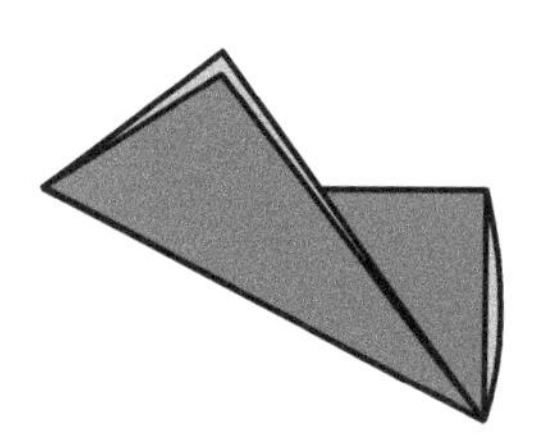

1 Fold in half bringing opposite edges together.

2 Fold the diagonal with both layers of paper.

3 Unit complete.

Assembly ★

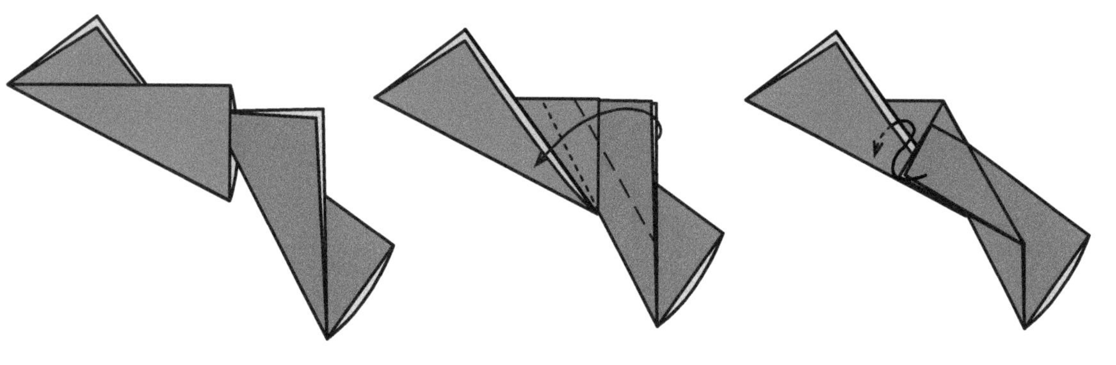

1 Insert the second unit into the pocket of the first unit.

2 Notice how the top edge of the second unit touches inside the pocket. Fold the raw edges of the second unit to lie on the diagonal of the first unit.

3 Tuck the layers of the second unit between the layers of the first unit.

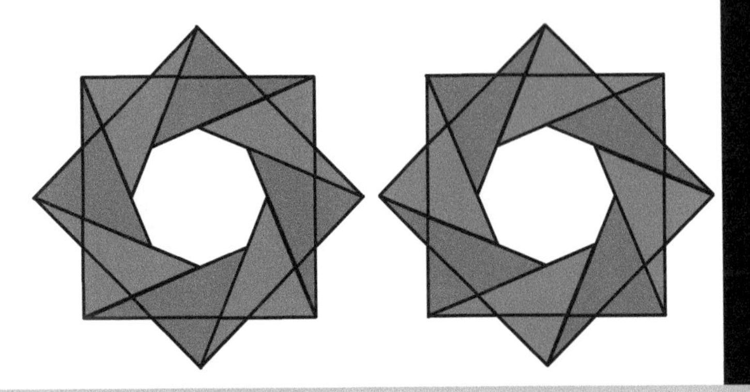

Handshake Wreath 8

★

Use eight 75 mm squares or larger

Finished model ratio 1.5

The paper and folding must be accurate for a stable assembly. The joining of units is not stable until all units are assembled. Even so, the model is stronger in tension than compression.

Module ★

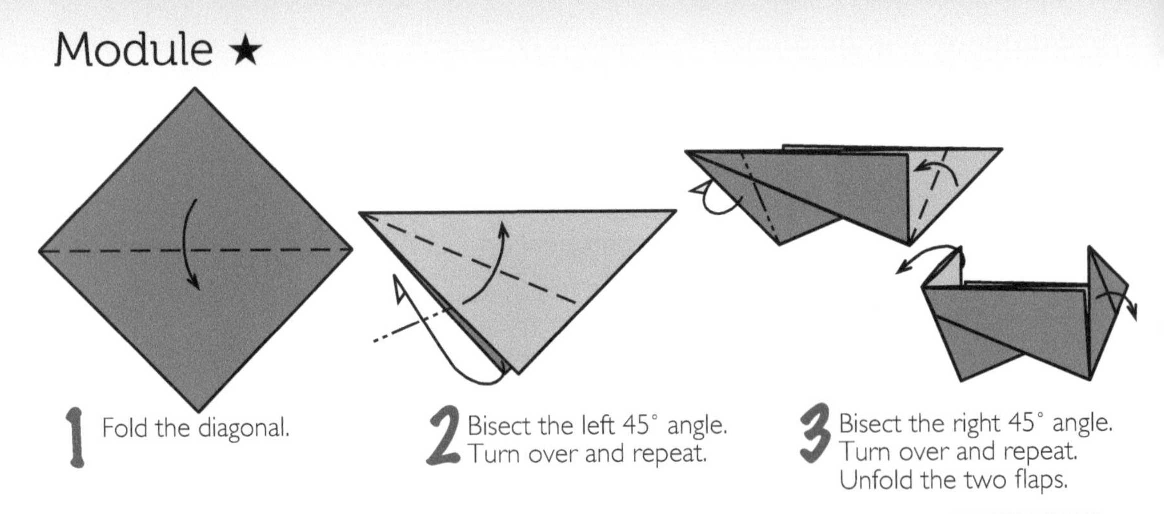

1 Fold the diagonal.

2 Bisect the left 45° angle. Turn over and repeat.

3 Bisect the right 45° angle. Turn over and repeat. Unfold the two flaps.

Assembly ★

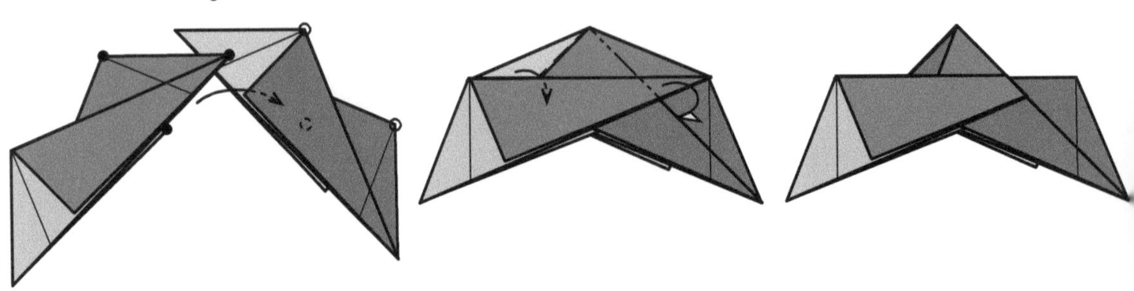

1 Insert the rear flap of the second between the top two layers of the first unit.

2 Fold the tips over to lock the units.

3 Continue adding units in a clockwise direction.

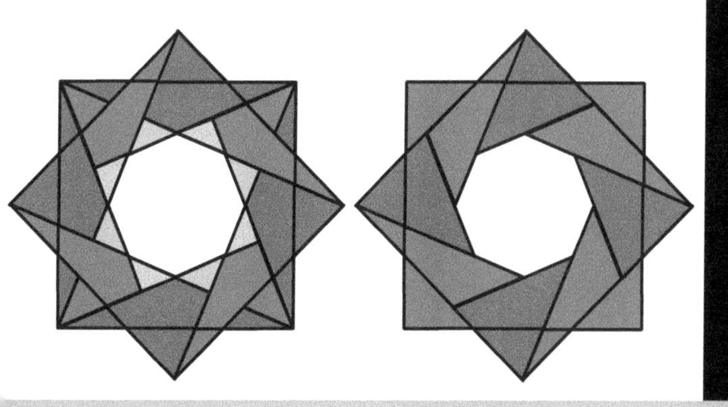

Hook & Roll Wreath 8

★★
Use eight 75 mm squares or larger
Finished model ratio 1.5

This variation of the previous model is more stable. During assembly, be sure to hold the units tight so that they do not move when you fold to lock.

Module ★★

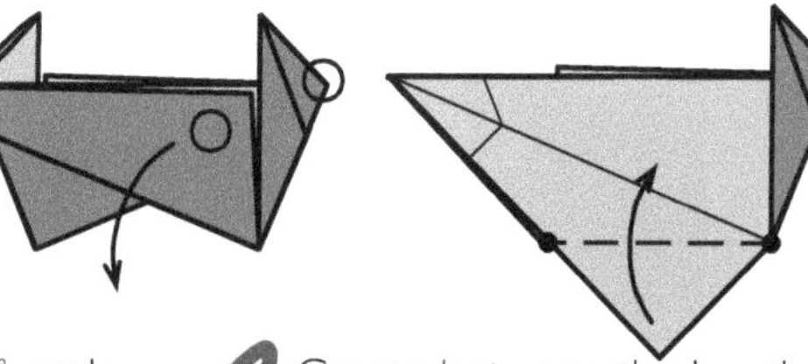

1 Start with a square folded on the diagonal. Bisect the left 45° angle. Turn over and repeat.

2 Bisect the right 45° angle. Turn over and repeat. Pull the corner of the square down to undo the left folds.

3 Crease between the location points. Do not repeat.

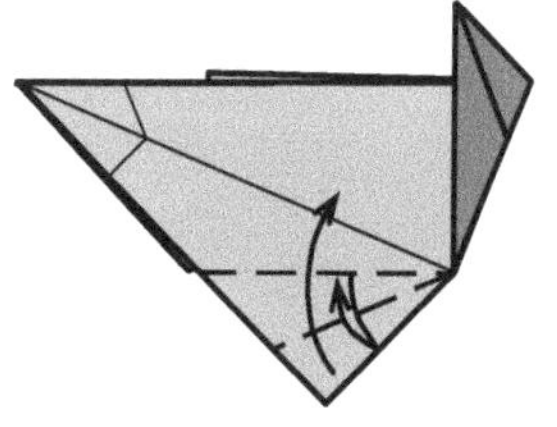

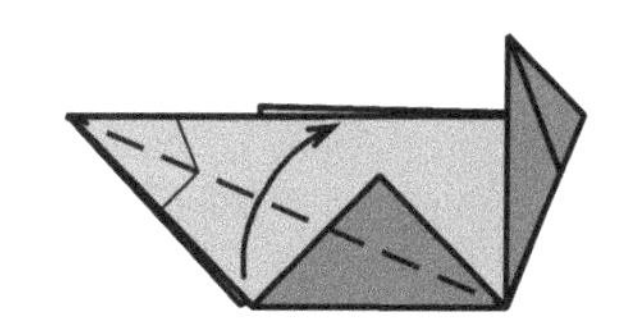

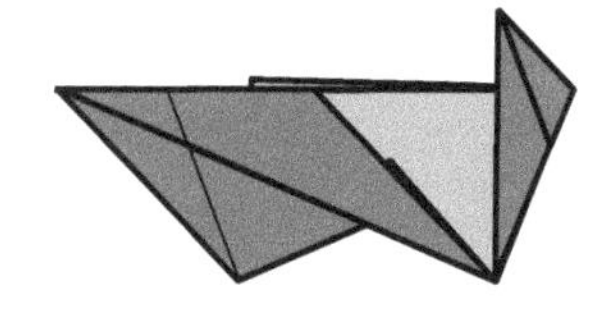

4 Bisect the right 45° angle, then refold the bottom flap.

5 For the first unit only, valley fold on the crease from step 1.

6 Only the first unit is like this. All other units are like step 5.

Assembly ★★

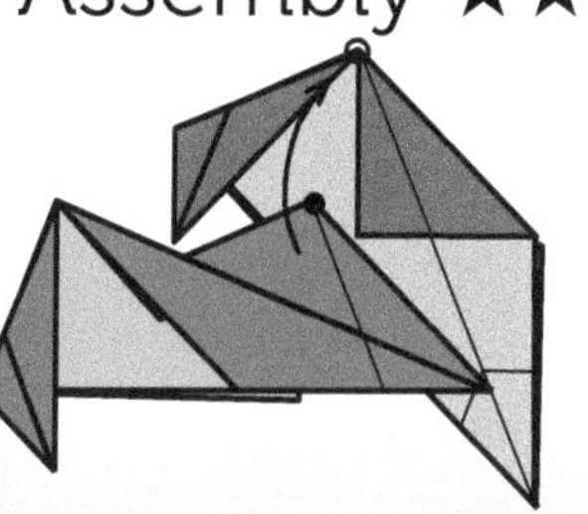

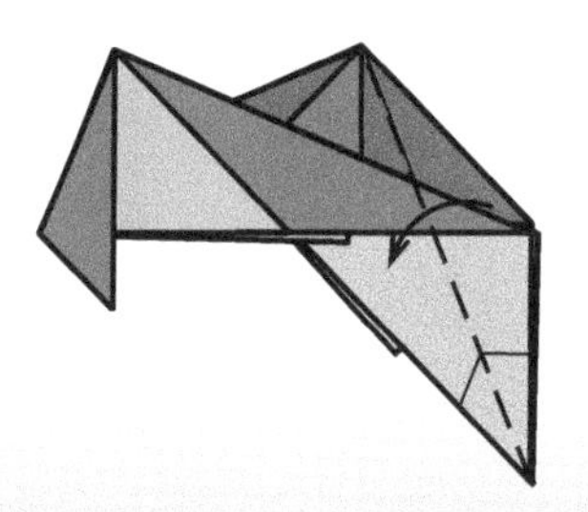

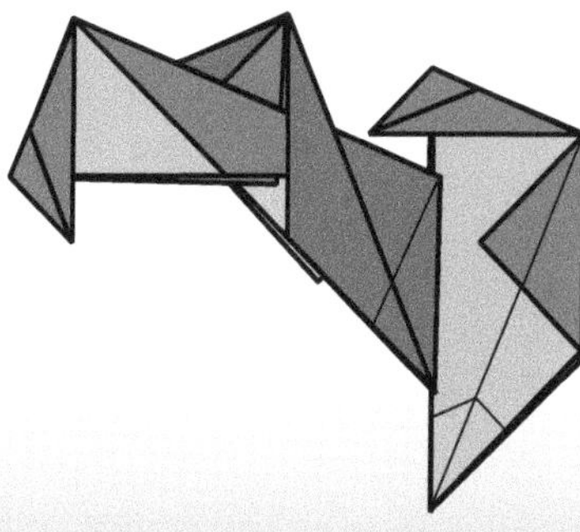

1 Hook the second unit onto the first unit.

2 Valley fold to lock.

3 Continue adding units in a clockwise direction.

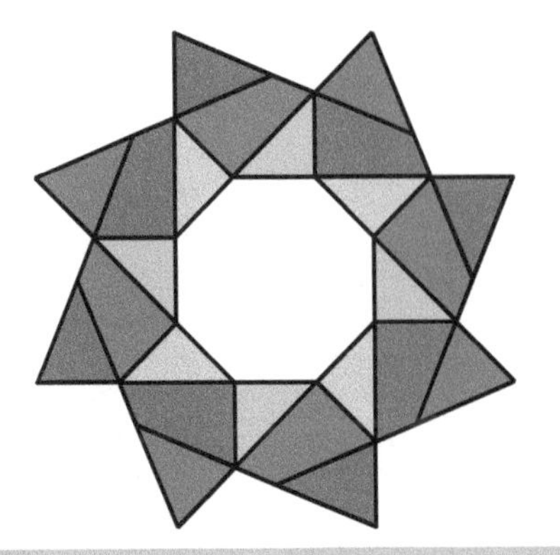

Star 8 Octad

★
Use eight 75 mm squares or larger
Finished model ratio 1.8

This model works well from paper that is white on one side and coloured on the other side. You can easily adapt the folding sequence for a different number of units and differently proportioned rectangles.

Module ★

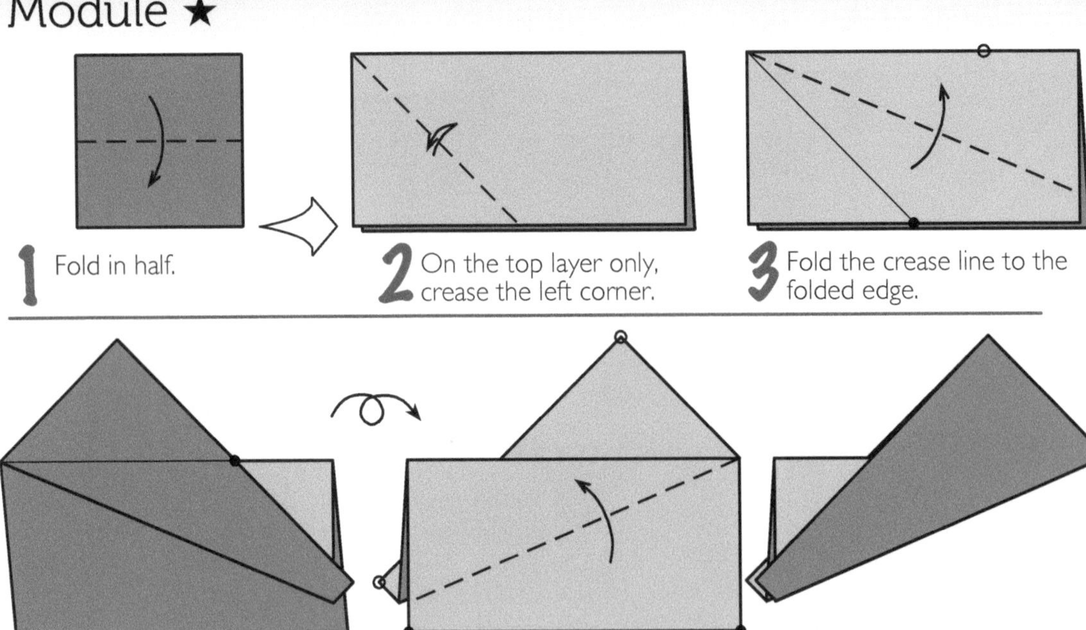

1 Fold in half.

2 On the top layer only, crease the left corner.

3 Fold the crease line to the folded edge.

4 Turn over.

5 Fold up to match the flap behind.

6 Unit complete.

Assembly ★

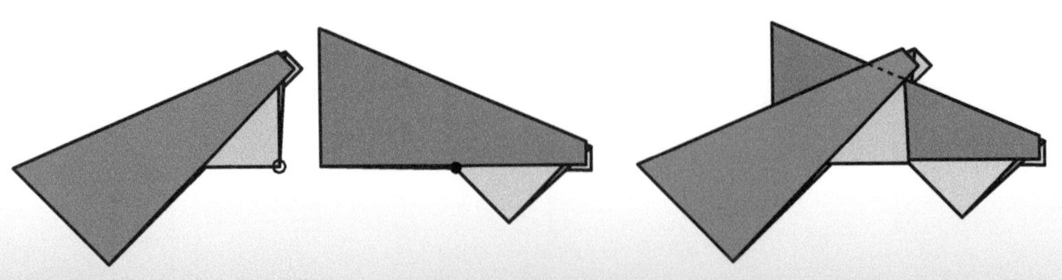

1 Insert the second unit into the first.

2 Fold the tips of the first unit inside the second to lock.

3 Continue adding units in a clockwise direction.

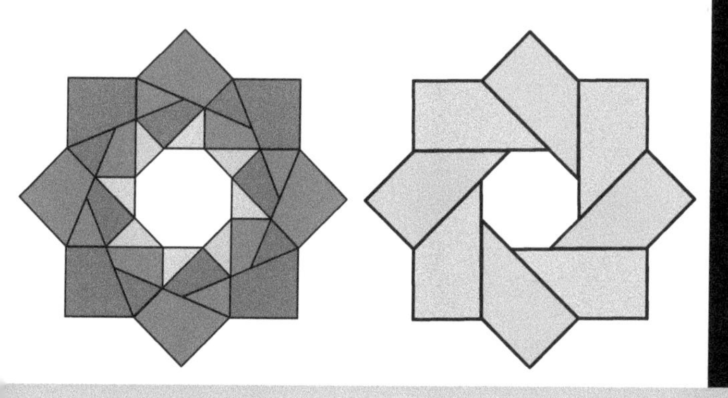

Starburst 8 Octad

★

Use eight 75 mm squares or larger

Finished model ratio 2.4

This model is closely related to the previous model. The two sides are different as it has no plane of symmetry.

Module ★

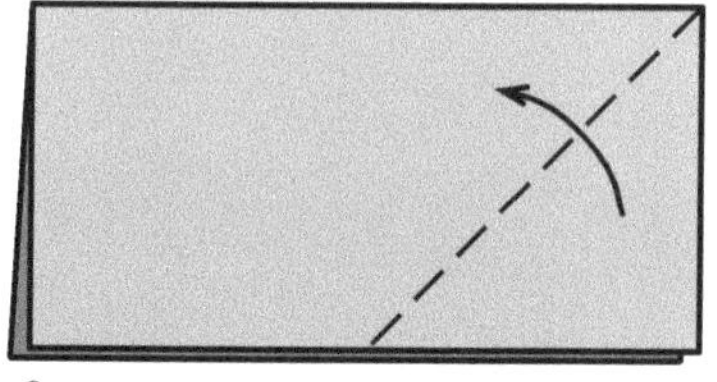

1 Start with a square folded in half. On the top layer only, fold the right corner.

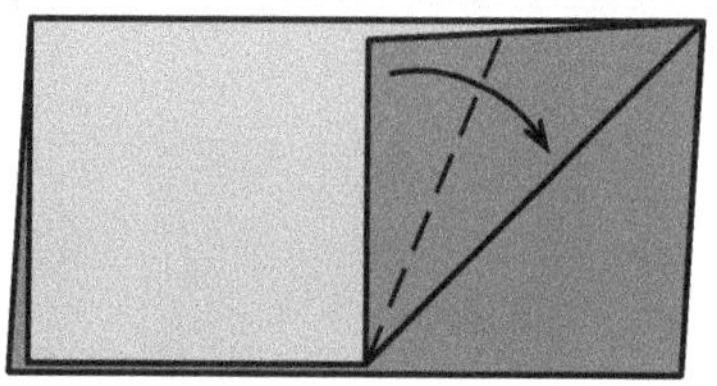

2 Bisect the 45° angle at the bottom of the flap.

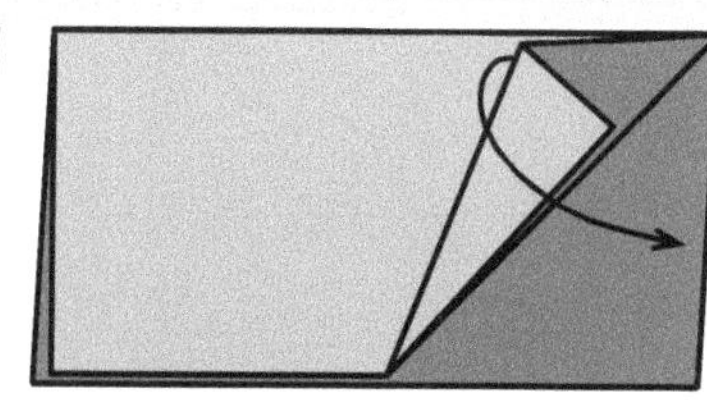

3 Fold the flap to the right and turn over.

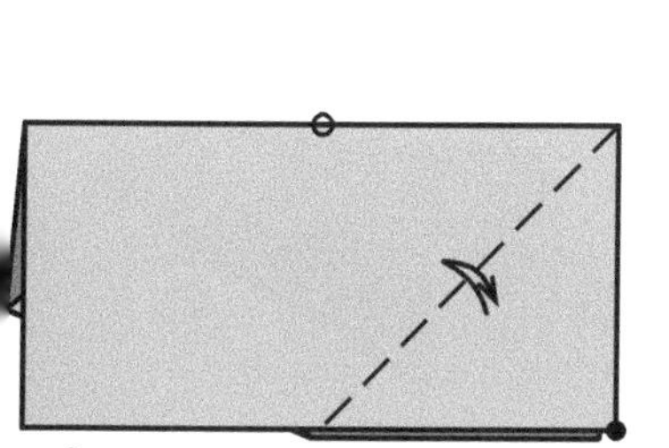

4 Fold the right corner and unfold.

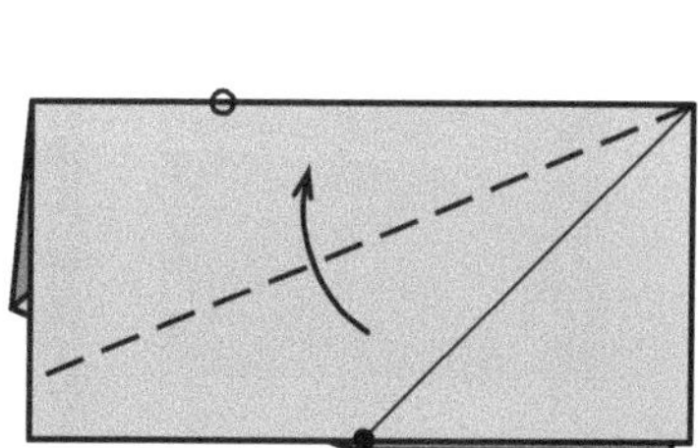

5 Fold the crease line to the folded edge.

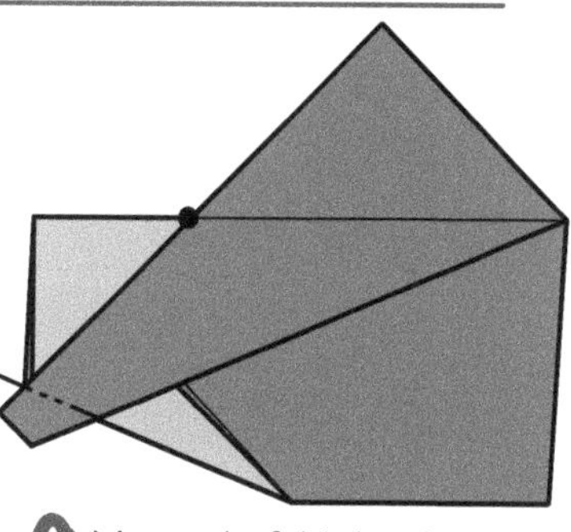

6 Mountain fold the tip and unfold.

Assembly ★

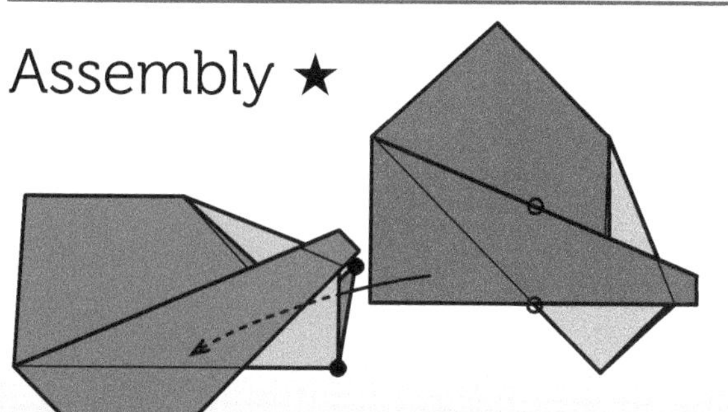

1 Insert the second unit into the first.

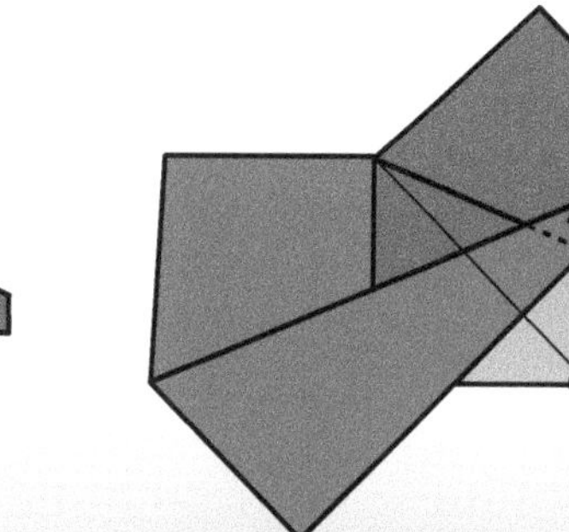

2 Fold the tips of the first unit inside the second to lock.

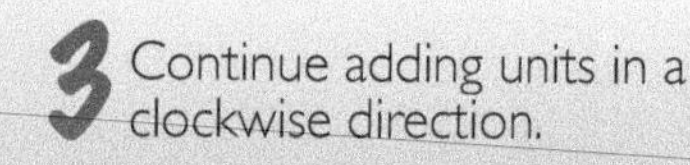

3 Continue adding units in a clockwise direction.

Starburst 8 60°

★

Use eight 75 mm squares or larger

Finished model ratio 2.3

Despite the 60° folds, eight units are required. When you have joined eight units, you can add extra locks on the other side by mountain folding the small flaps into the pockets behind.

Module ★★

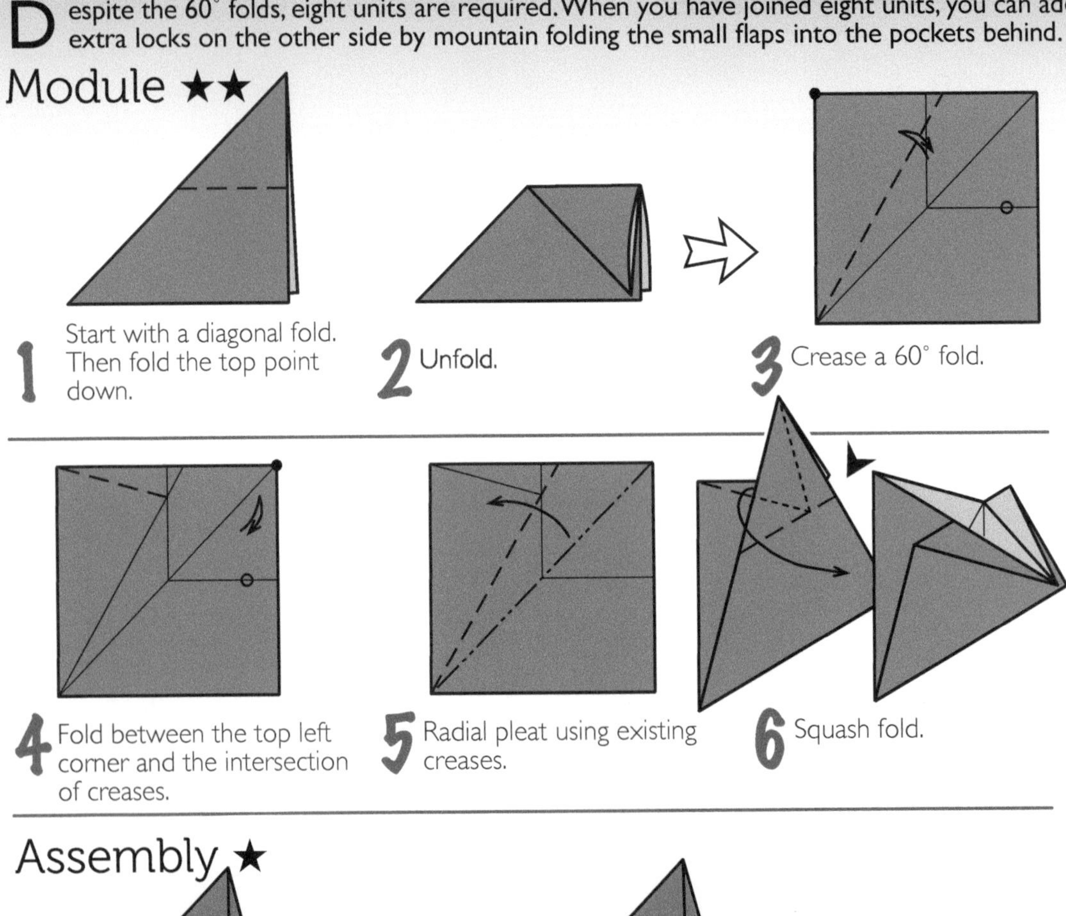

1 Start with a diagonal fold. Then fold the top point down.

2 Unfold.

3 Crease a 60° fold.

4 Fold between the top left corner and the intersection of creases.

5 Radial pleat using existing creases.

6 Squash fold.

Assembly ★

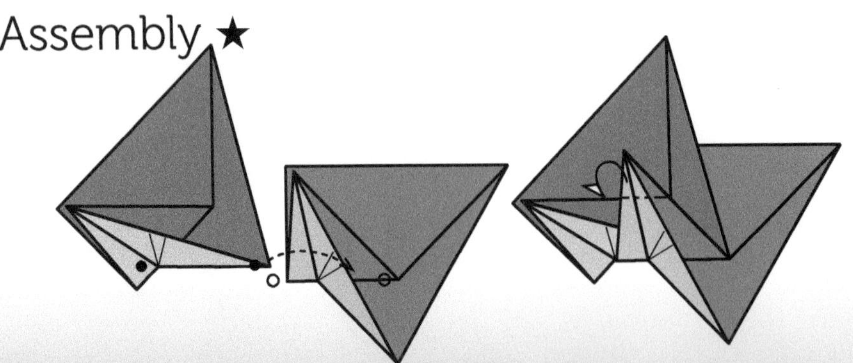

1 Insert the second unit into the first.

2 Fold the tip of the first unit behind to lock.

3 Continue adding units in an anticlockwise direction.

Starburst 6 Hexad

★

Use six 75 mm squares or larger

Finished model ratio 2.2

The units could be folded from double bronze rectangles but the extra paper in the square adds interest to the unit and final appearance.

Module ★

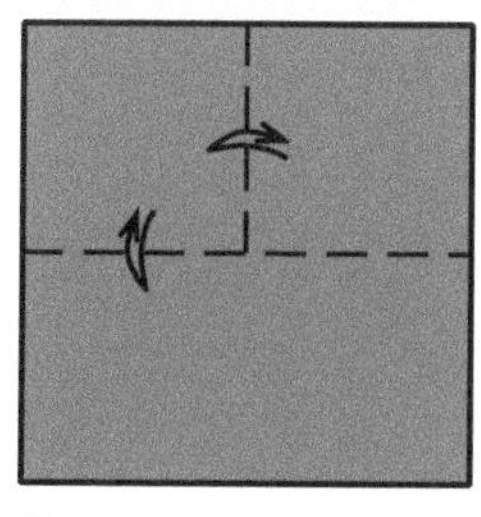

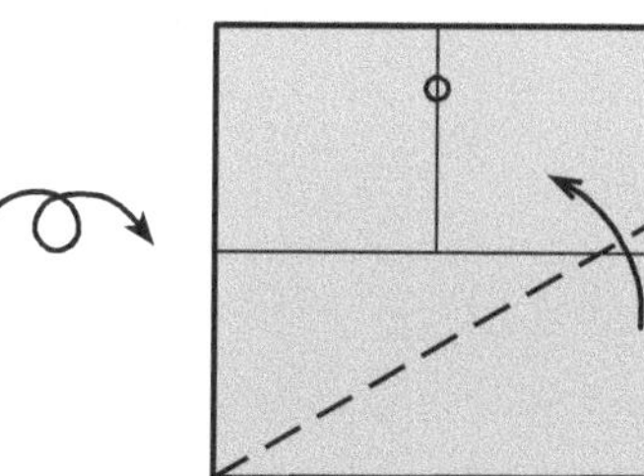

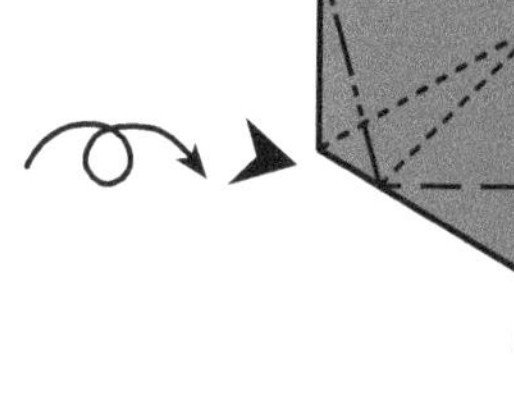

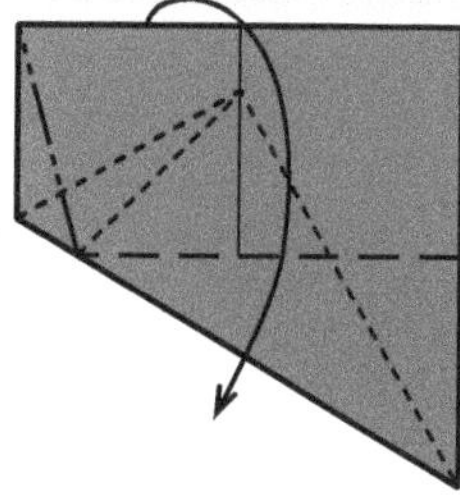

1 Crease a 2 by 1 grid in the top half. Turn over.

2 Make a 60° fold. Turn over.

3 Fold the top part down and squash the left flap.

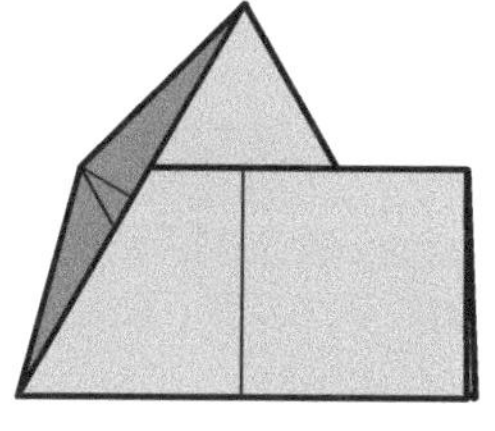

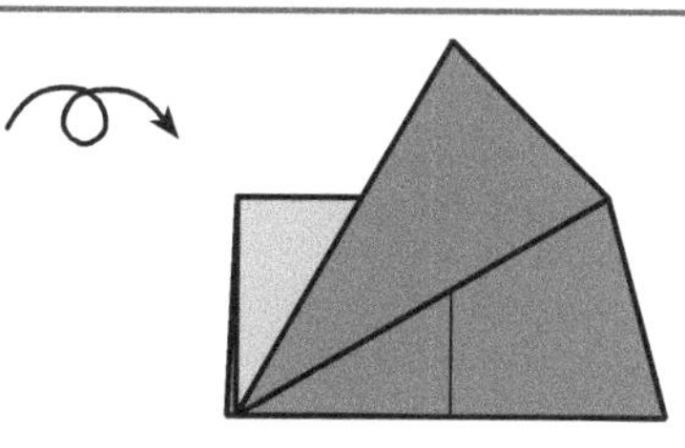

4 Turn over.

5 Unit completed.

Assembly ★

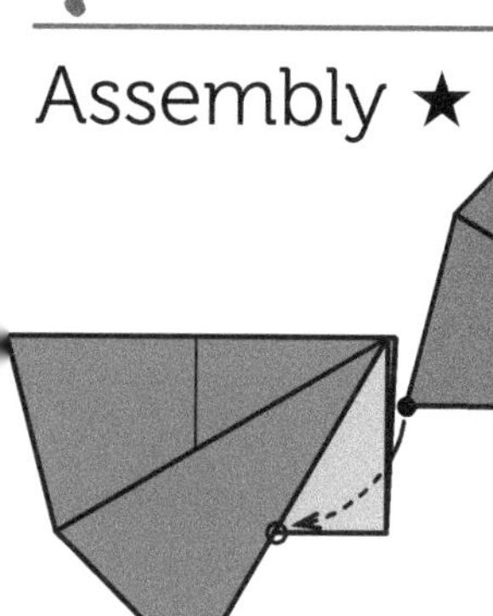

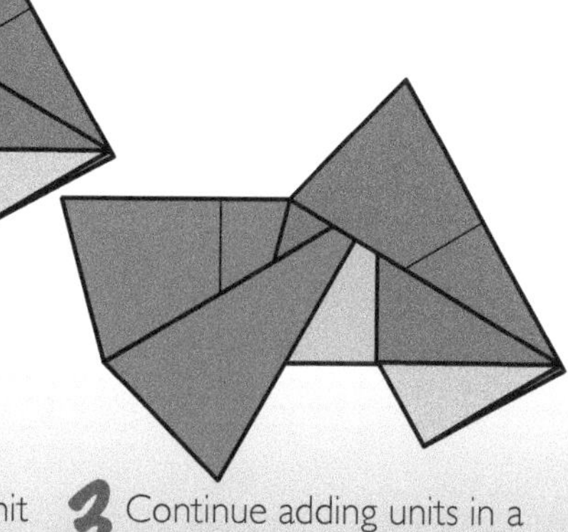

1 Insert the second unit into the first.

2 Fold the tips of the first unit inside the second to lock.

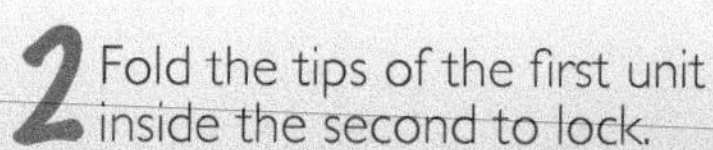

3 Continue adding units in a clockwise direction.

Chapter 9

Stars

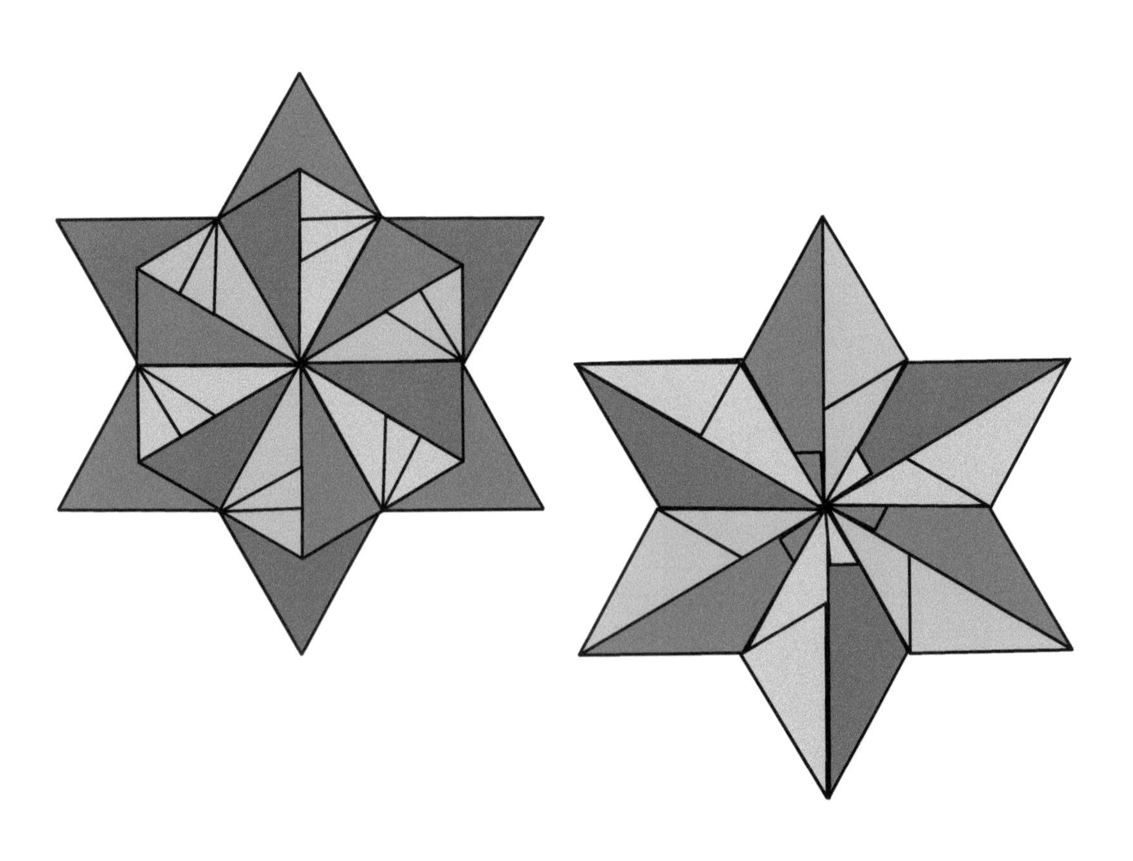

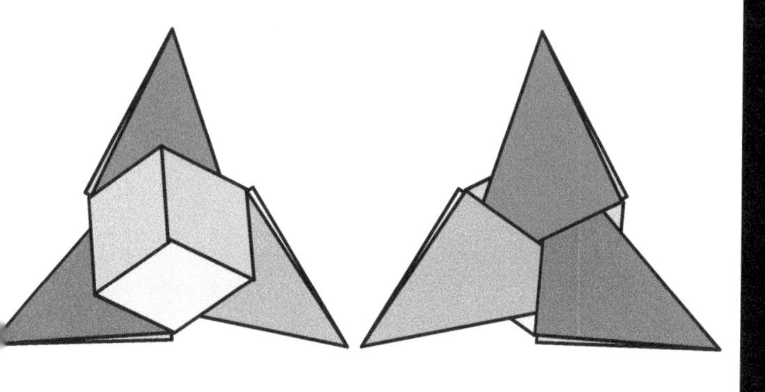

Star 3

Square From Silver Rectangle

★

Use three silver rectangles A8 or larger and one for the template

Finished model ratio 2.0

Using a template for the 60° folds avoids extraneous creases and can be more accurate. The model relies on √2+1 being slightly greater than 4√3÷3, i.e. 2.414 > 2.309. Shorter rectangles would make a bigger central hole.

Template ★

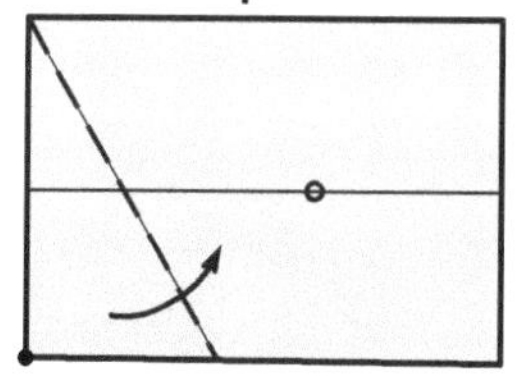

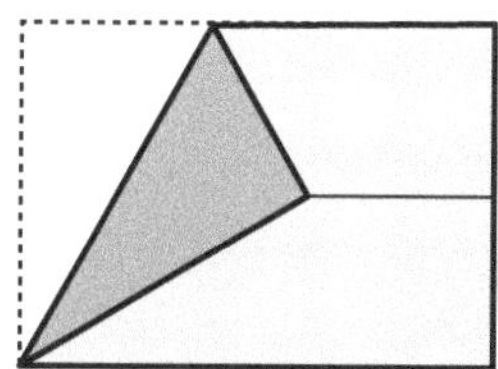

Module ★

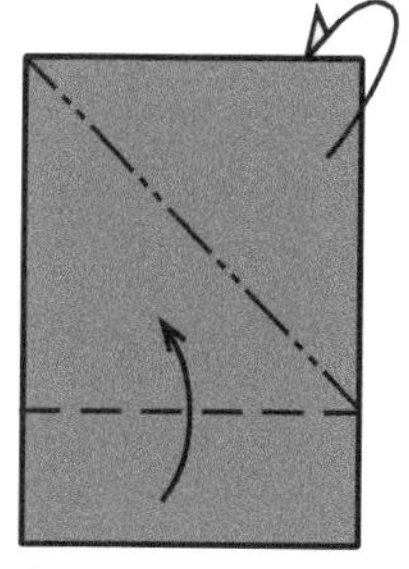

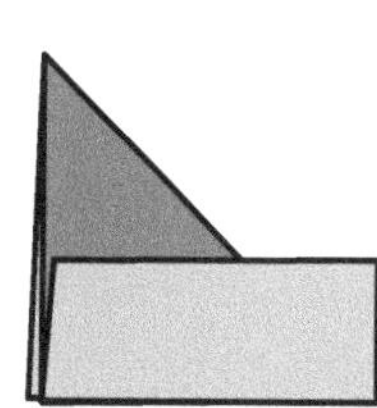

1 Perform the square from rectangle procedure.

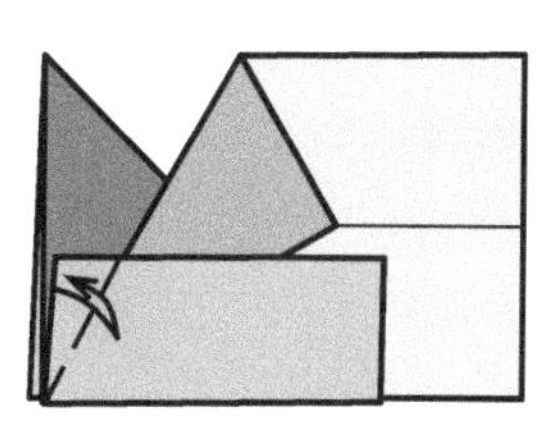

2 Insert the template and valley fold the left corner.

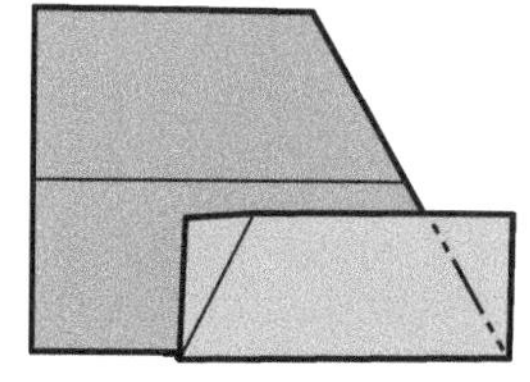

3 Remove the template and insert. Mountain fold the right corner.

Assembly ★

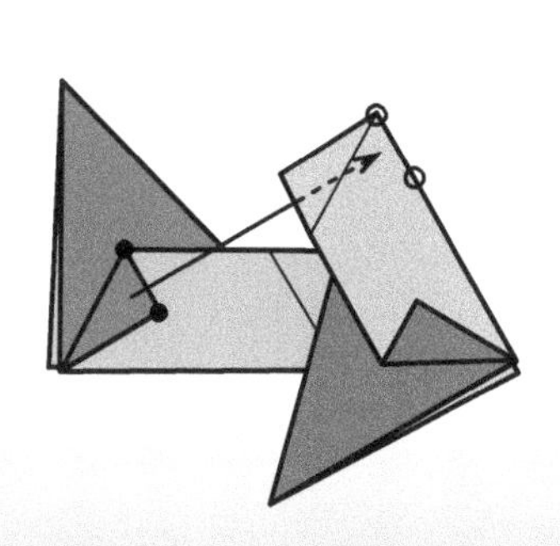

1 Hook the first unit into the second unit.

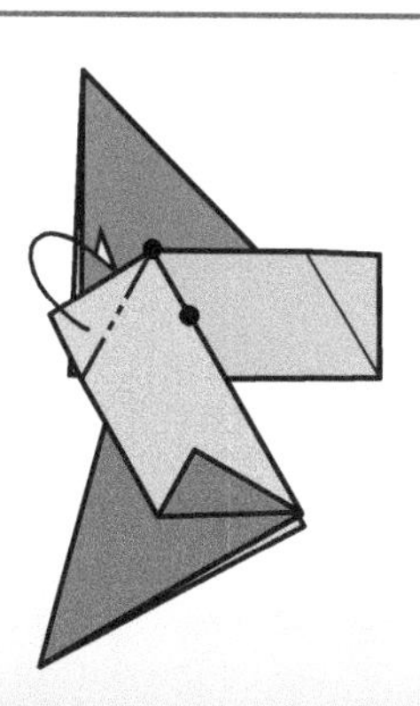

2 Mountain fold behind to lock.

3 Add the third unit to the second unit. Join it to the first.

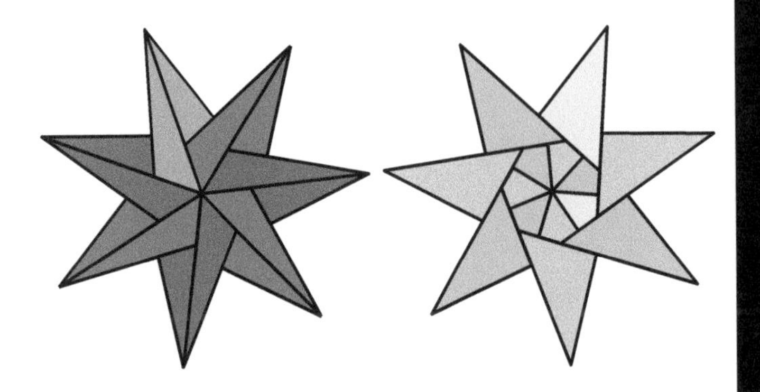

Silver Star 7

★★

Use seven A7 rectangles

Finished model ratio 2.5

An accurate bisection in step 4 would make the angle of 51.0576°, close to the central angle of a regular heptagon. However, folding a little further means that the flap fits the pocket behind without paper being crushed.

Module ★★

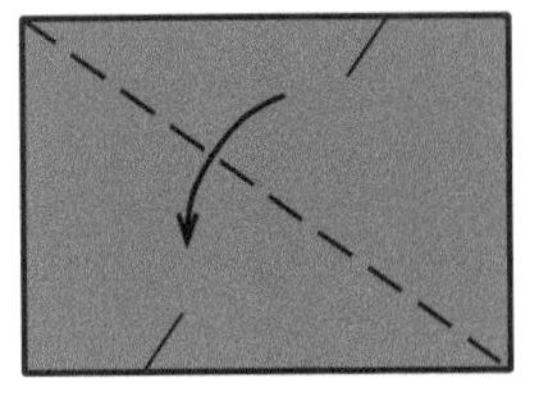

1 Fold the diagonal.

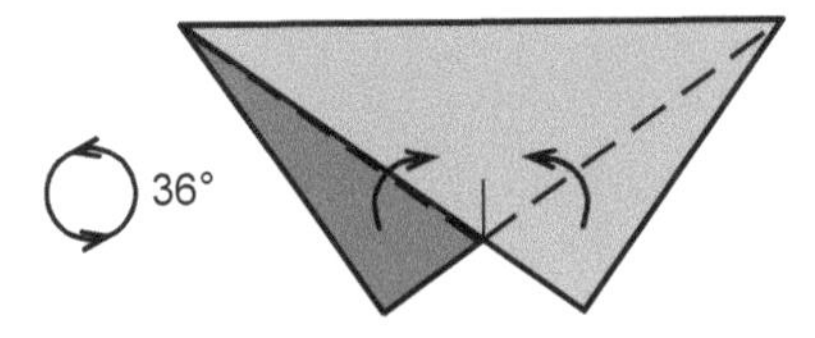

2 Fold left then right flaps.

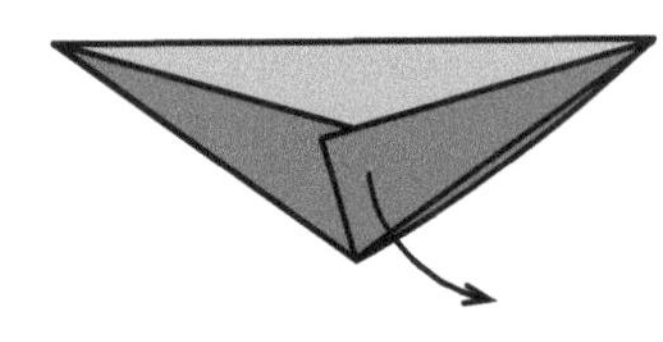

3 Pull the right flap out.

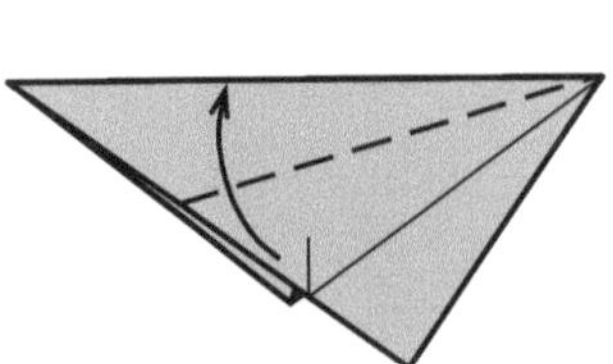

4 Fold so that the crease line is a bit above the folded edge.

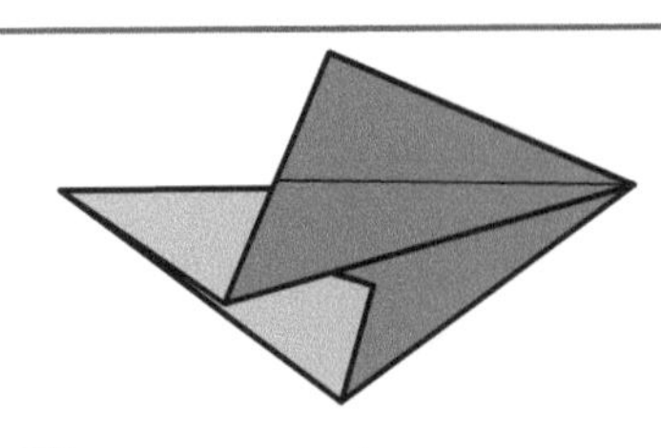

5 Unit complete.

Assembly ★★

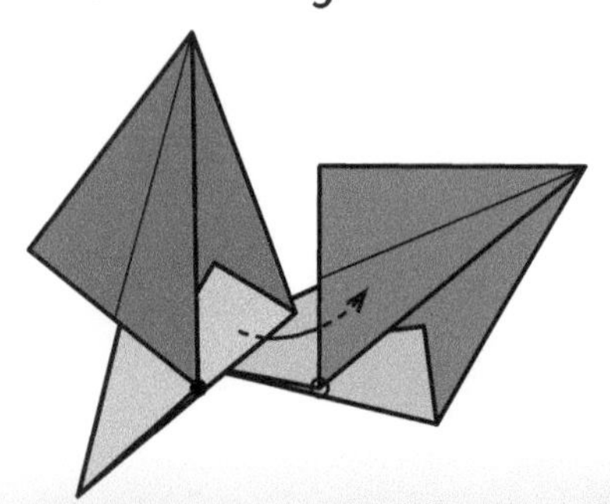

1 Insert the second unit into the first unit. The folded edges meet.

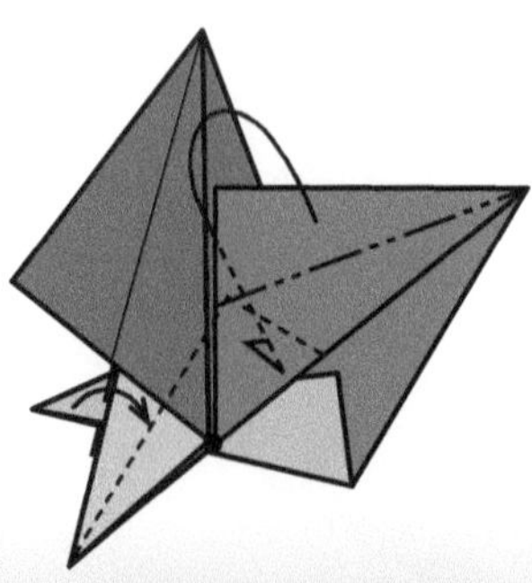

2 Tuck the flap of the first unit behind and into the pocket of the second unit. Fold the little tip over to lock.

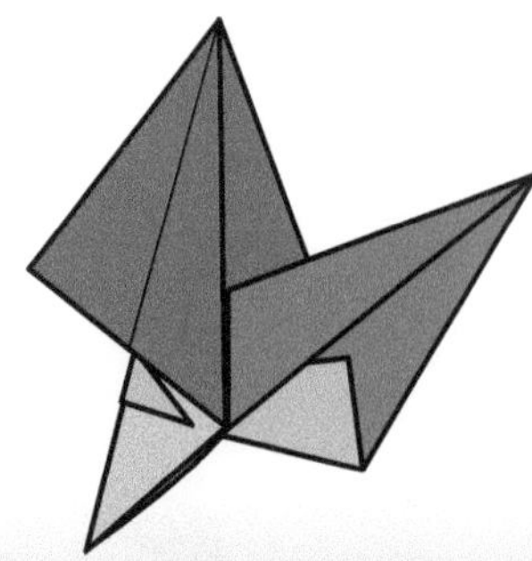

3 Continue adding units in an anticlockwise direction.

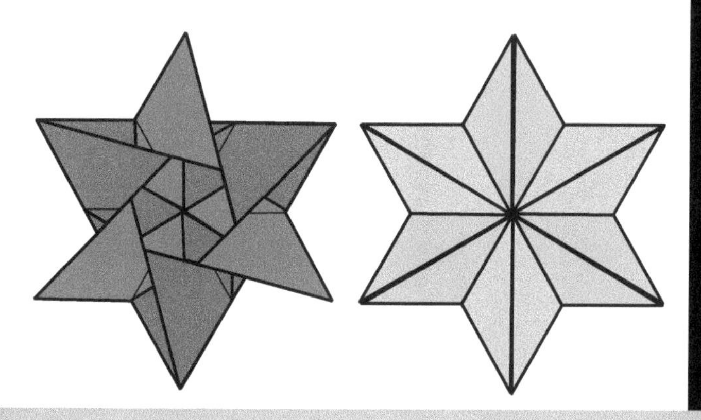

Star 6

★
Use six 75 mm squares
Finished model ratio 2.0

This model is like the previous model but uses 60° geometry.

Module ★

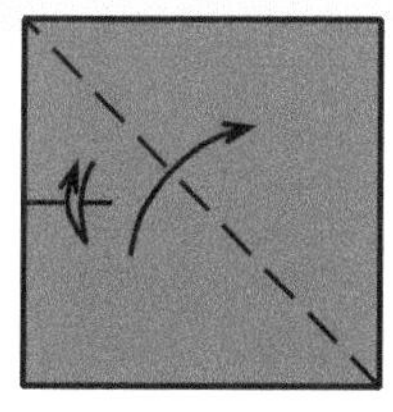

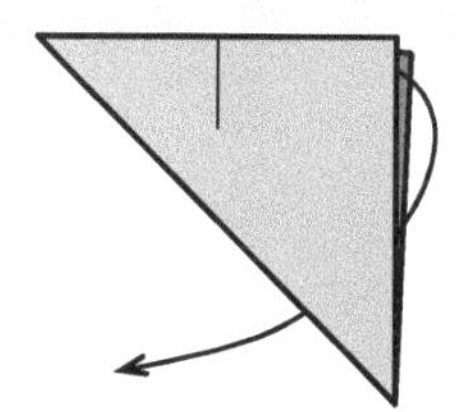

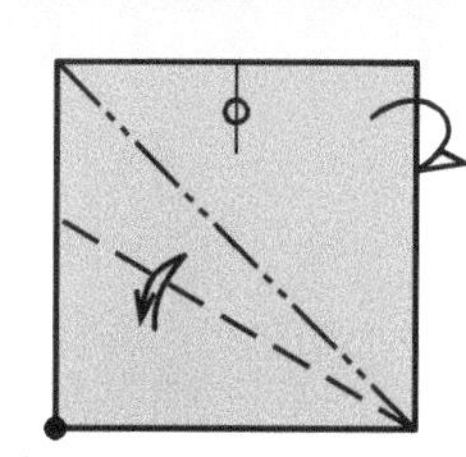

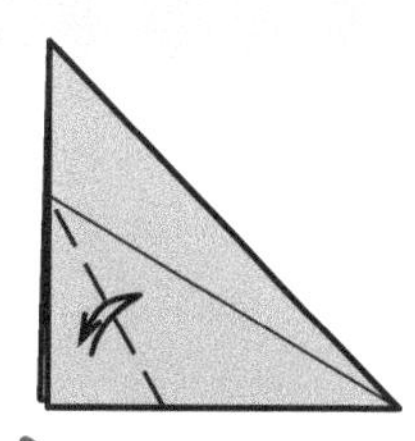

1 Pinch midpoint on left edge. Fold the bottom left corner to top right corner.

2 Unfold from behind.

3 Crease a 60° fold, then refold the diagonal.

4 Bisect 60°.

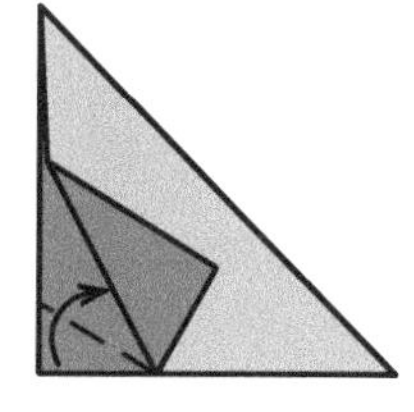

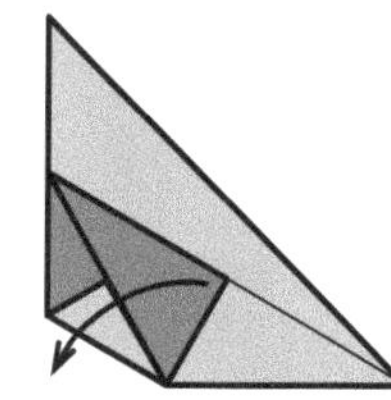

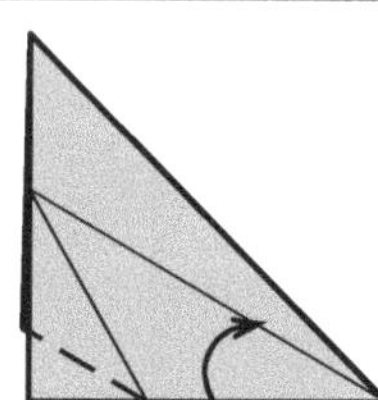

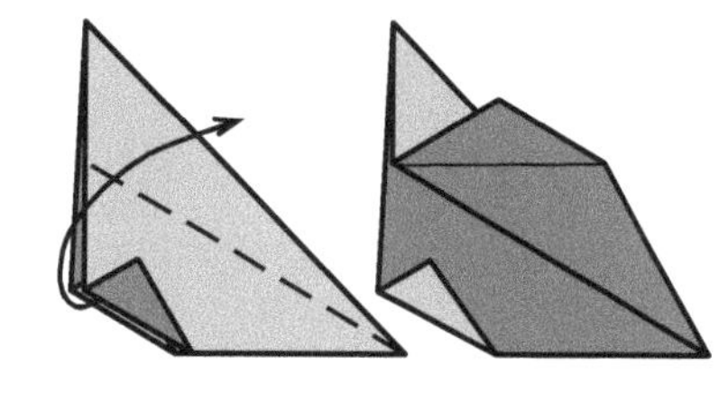

5 Fold to meet the folded edge.

6 Unfold.

7 Fold the top layer to match.

8 Fold up.

Assembly ★

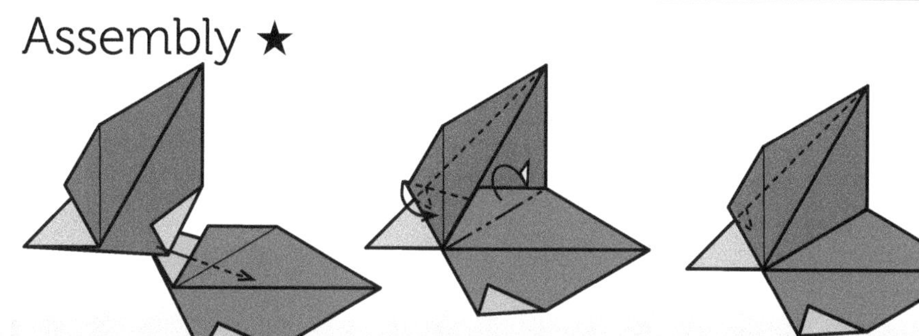

1 Slide the second into the first.

2 Mountain fold the flap of the first unit into the second.

3 Fold the tip of the first unit lock.

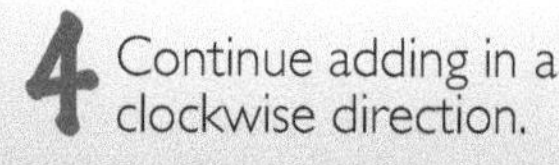

4 Continue adding in a clockwise direction.

Pentagram

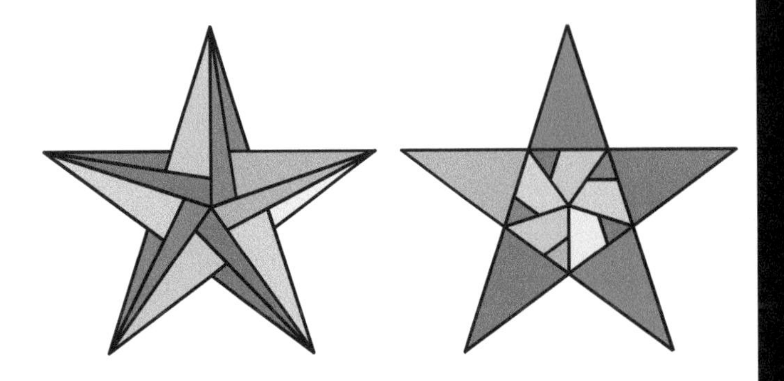

★★
Use five 100 mm squares
Finished model ratio 1.9

An earlier version of this model used the 1:3 approximation for dividing a right angle into fifths. However, this version uses a more accurate method which is important for the little locking flaps. For extra accuracy, with larger paper, the midpoint lands just below the diagonal in step 5.

Module ★★

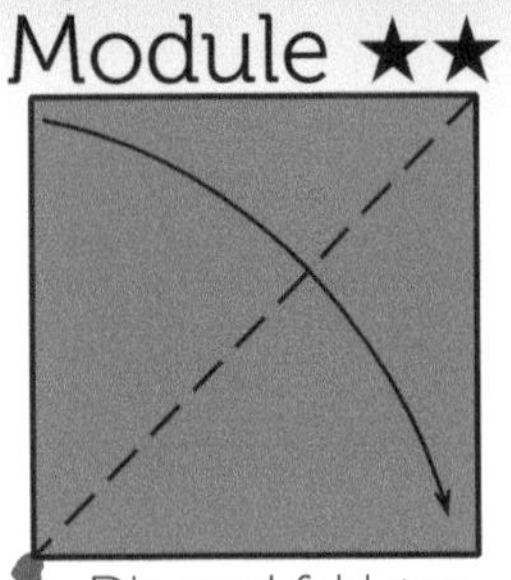

1 Diagonal fold, top left to bottom right corner.

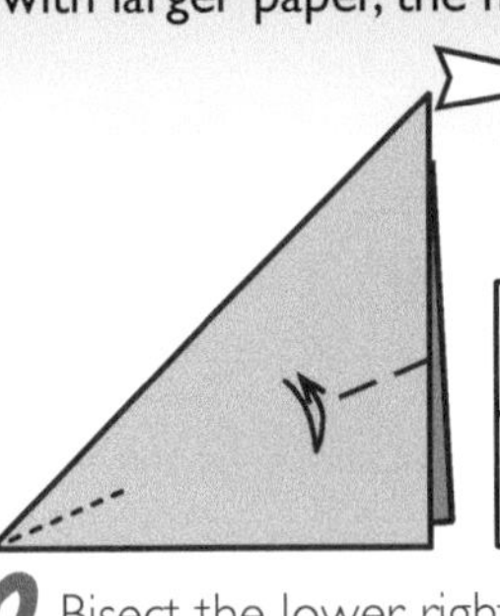

2 Bisect the lower right 45° angle but do not fold all the way: only pinch the right edge.

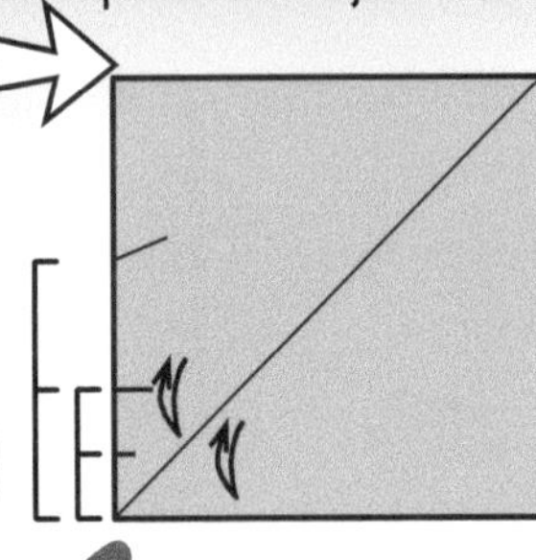

3 Pinch halfway up to the pinch from step 2. Then repeat with the new pinch.

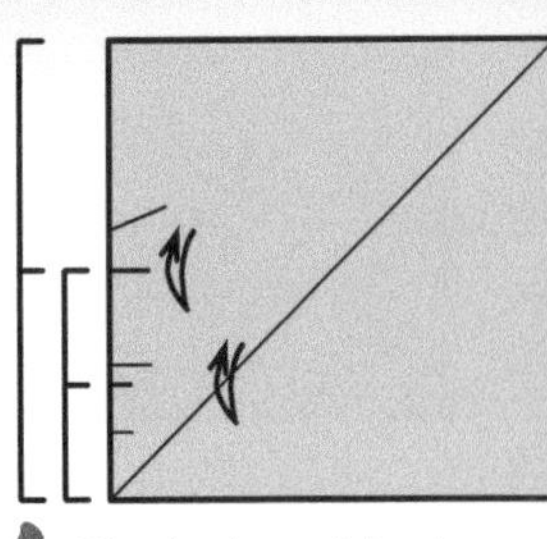

4 Pinch the midpoint of the left edge and then quarter point.

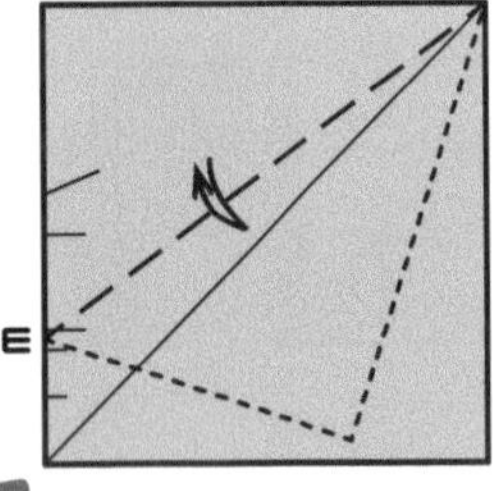

5 Crease through the top right corner and the midpoint of the two closest pinches.

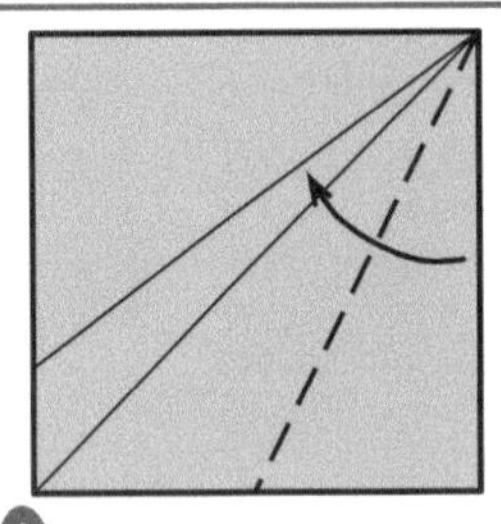

6 Fold up to the crease.

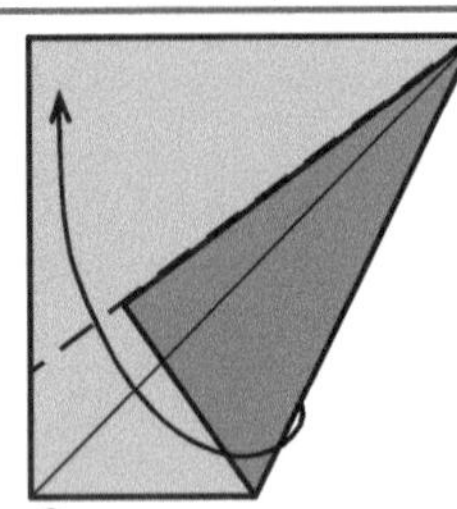

7 Fold over.

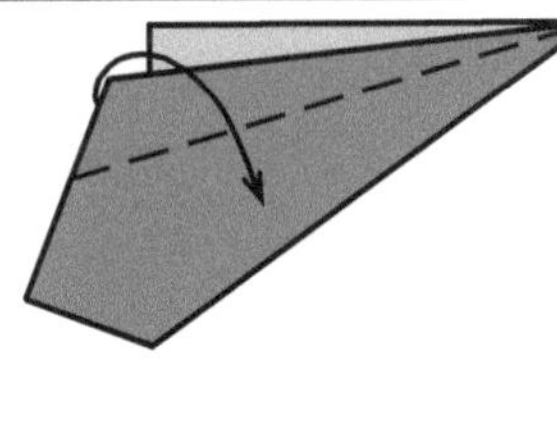

8 Fold onto the diagonal.

Assembly ★★★

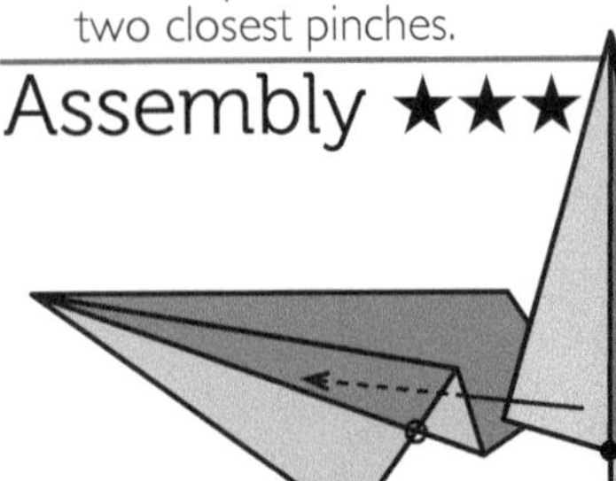

1 Slide the second unit into the first.

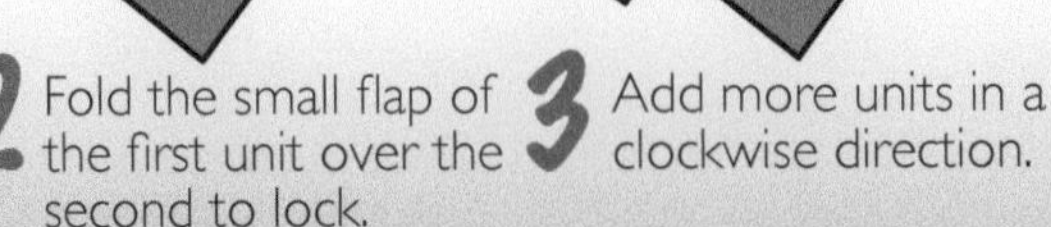

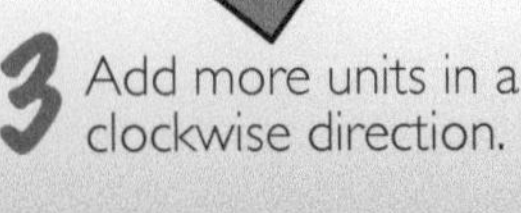

2 Fold the small flap of the first unit over the second to lock.

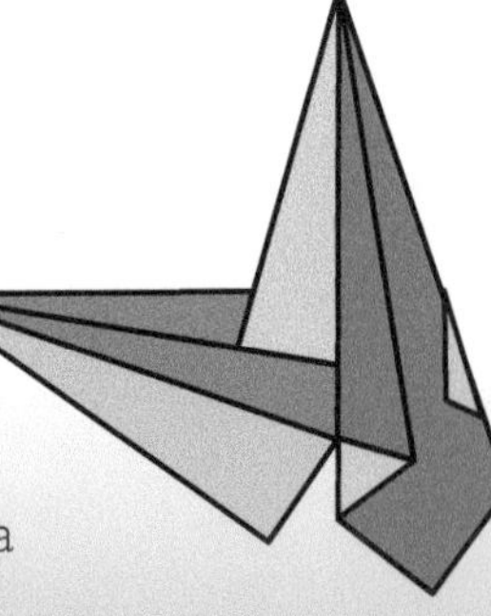

3 Add more units in a clockwise direction.

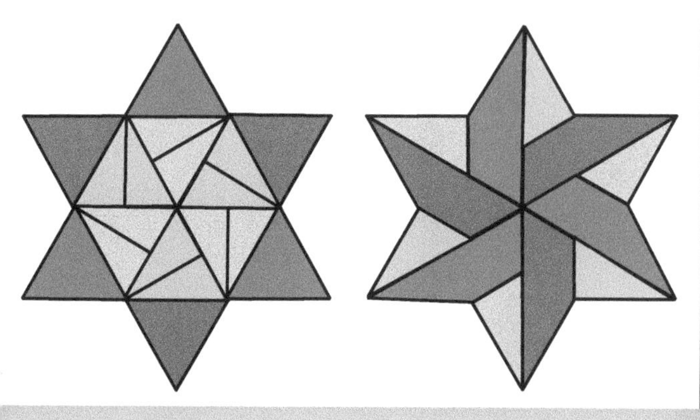

Star 6
Bronze Rectangle From Square

★★
Use six 75 mm squares or larger
Finished model ratio 1.7

This unit could be folded from a double bronze rectangle but can be efficiently made from a square. This model is a 60° version of Star 4, Square from Rectangle.

Module ★★

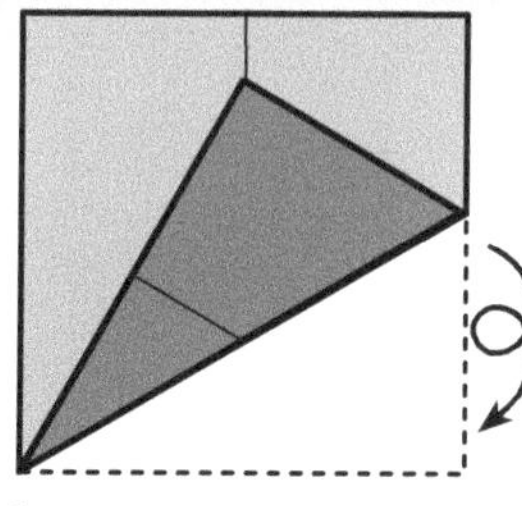

1 Make a 60° fold. Turn over.

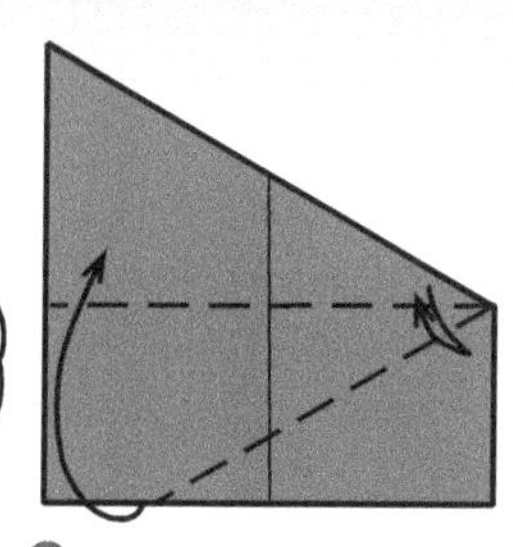

2 Bisect the 120° angle and unfold. Then fold bottom flap up.

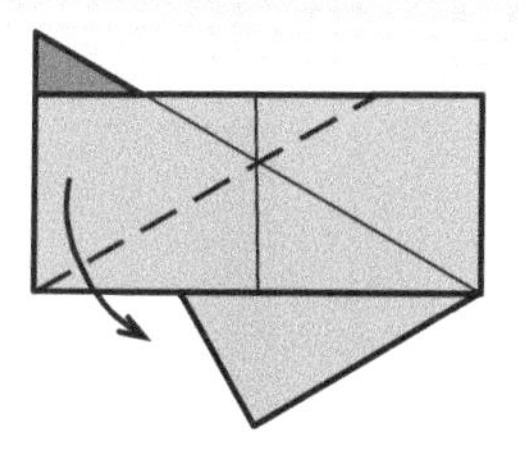

3 Bisect 120°, folding through the intersection of creases.

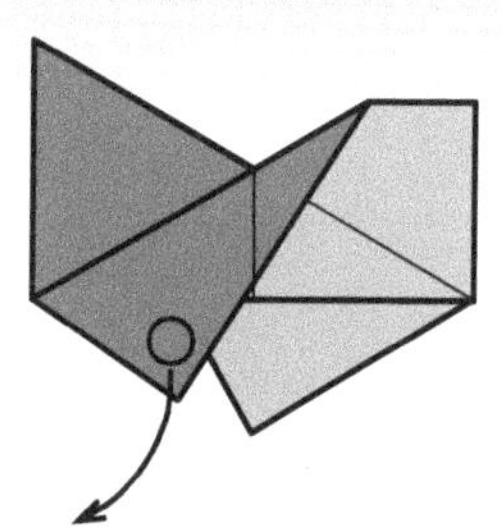

4 Pull the bottom left flap downwards.

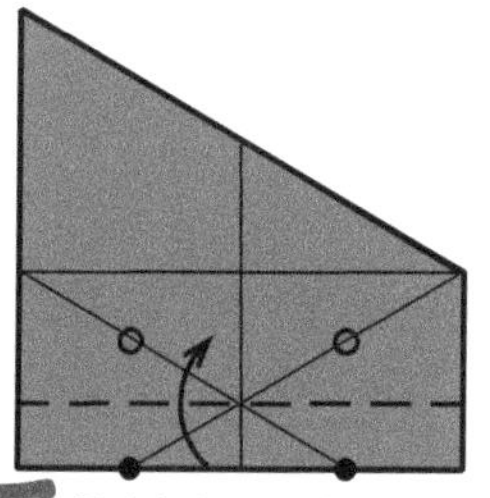

5 Fold through the intersection of creases.

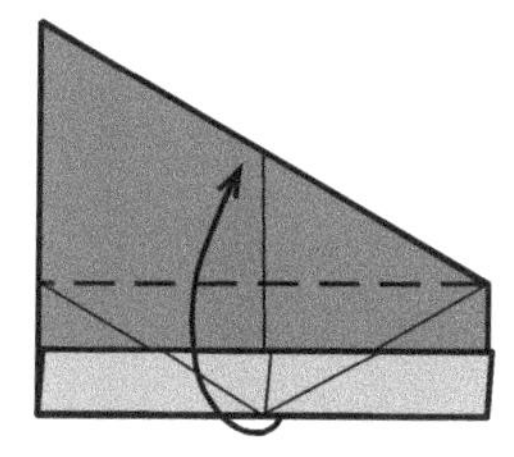

6 Fold up on the existing crease.

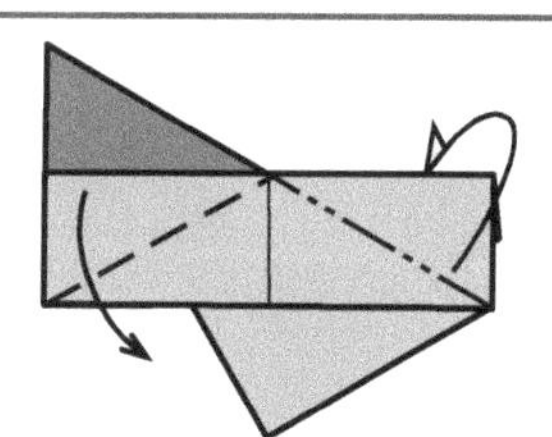

7 Fold the left flap down and the right flap behind.

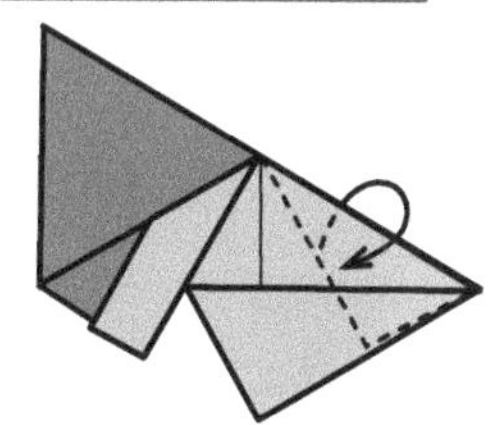

8 Unfold the right flap.

Assembly ★

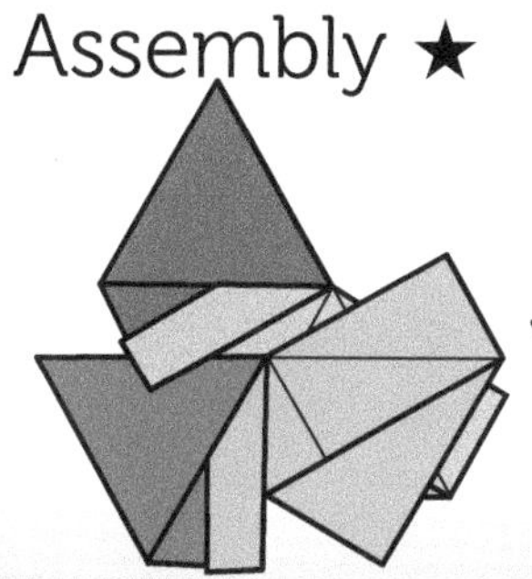

1 Hook the second unit onto the first and slide right.

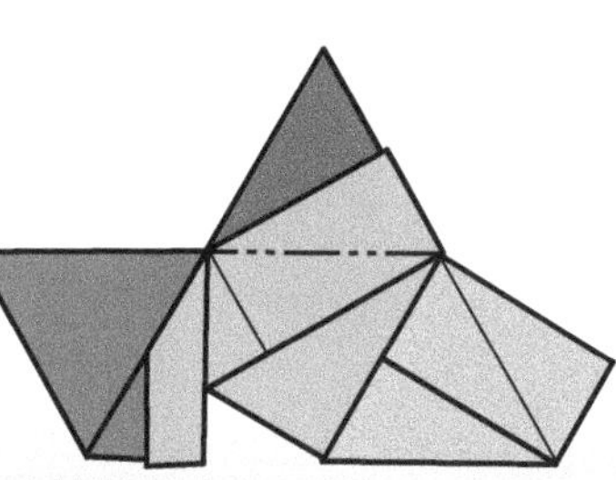

2 Tuck the flap of the second unit behind.

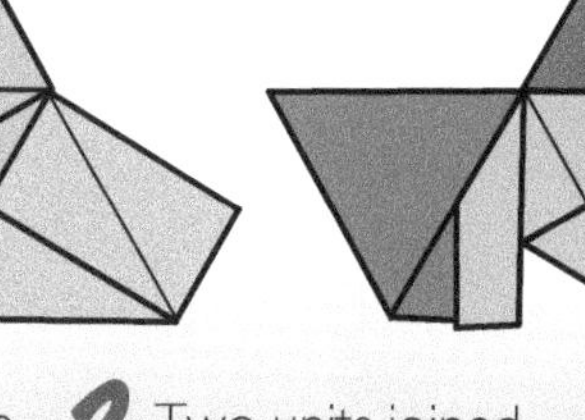

3 Two units joined.

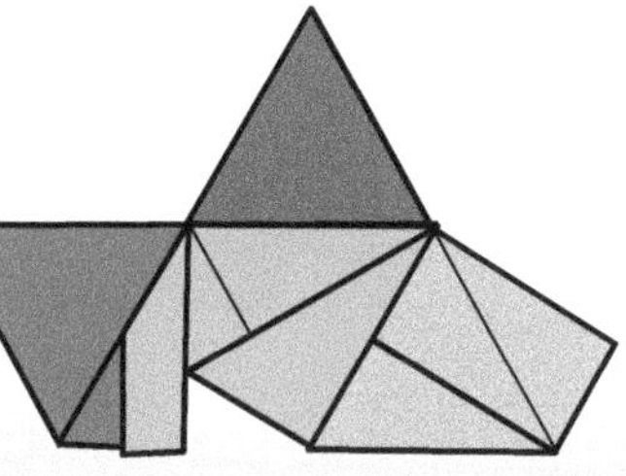

4 Continue adding units in a clockwise direction.

Handshake Star 6

★

Use six 4:√3 rectangles
Finished model ratio 1.7

You can make four 4:√3 rectangles from a bronze rectangle, or two from a square with little waste.

Module ★

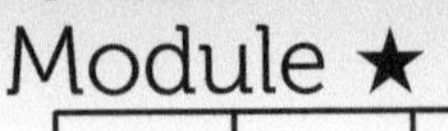

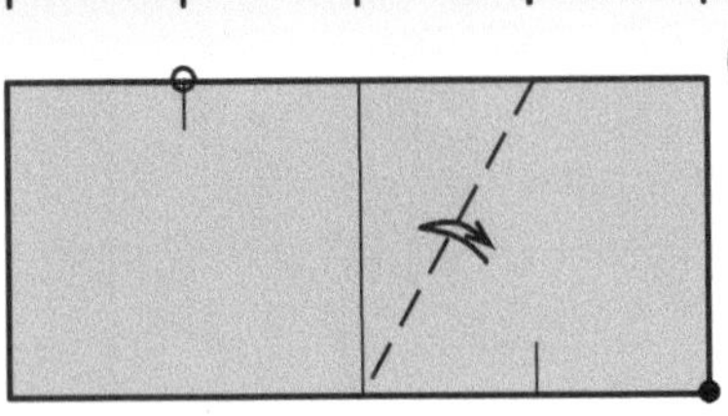

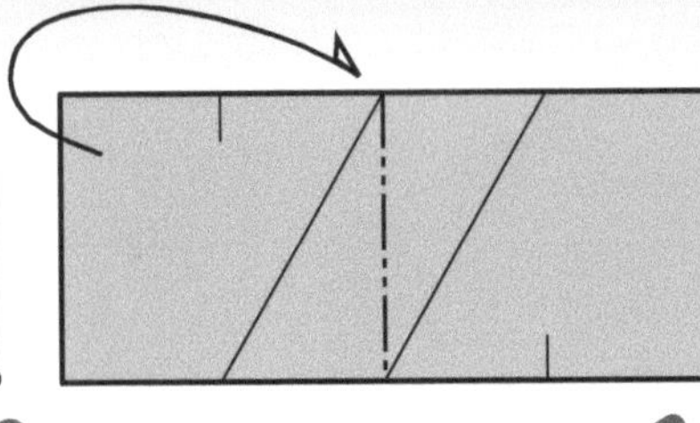

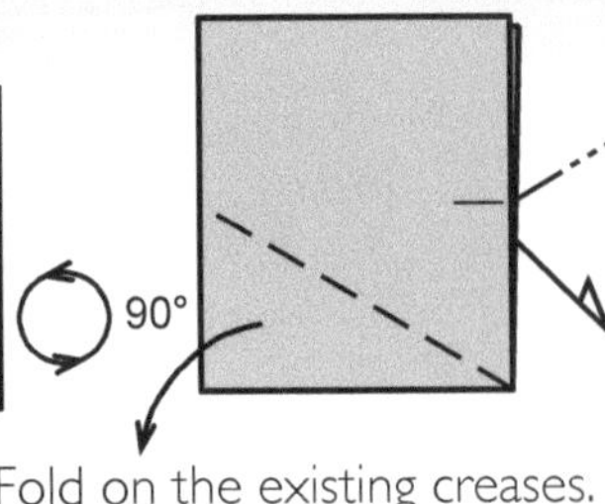

1 Fold the lower right corner to a point one quarter along the top edge and unfold. Rotate 180° and repeat.

2 Mountain fold in half behind. Rotate 90° anticlockwise.

3 Fold on the existing creases.

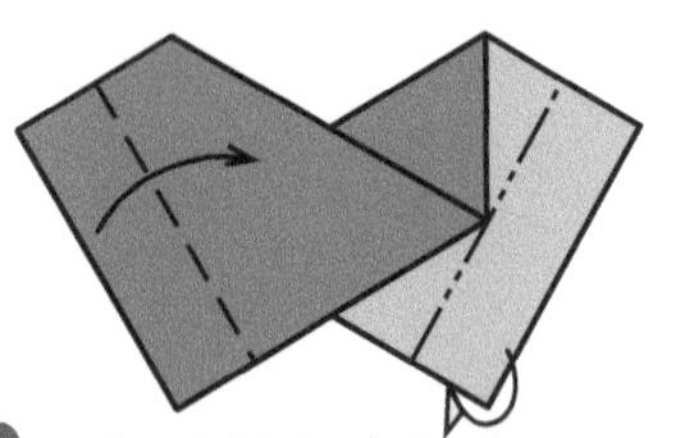

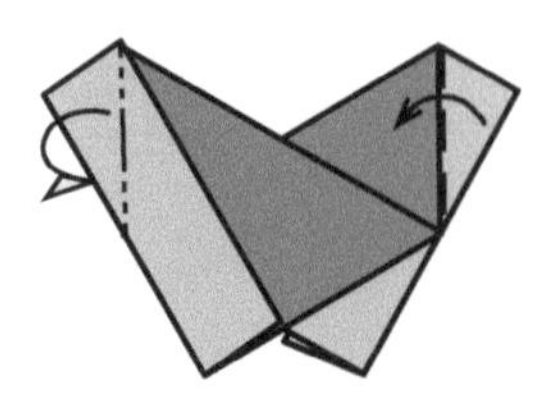

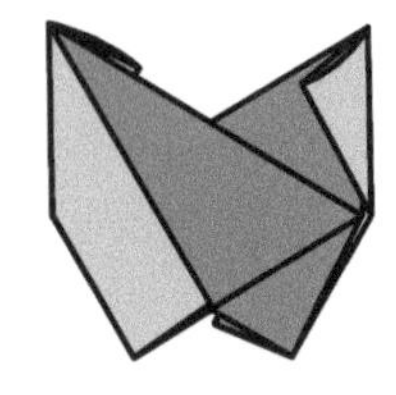

4 Valley fold the left edge. Turn over and repeat.

5 Fold the right tip over the raw edge. Turn over and repeat.

6 Unfold the tips.

Assembly ★

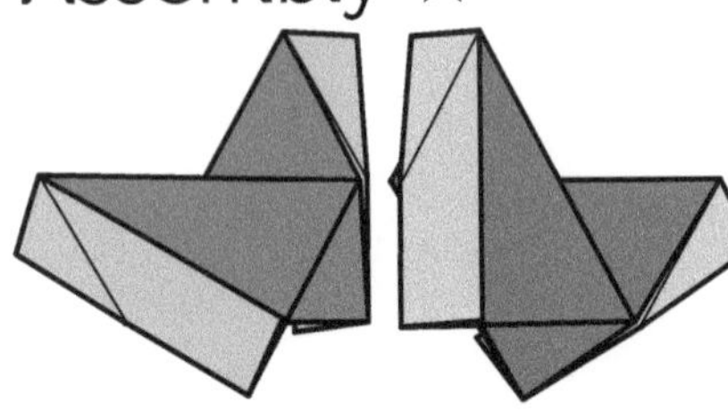

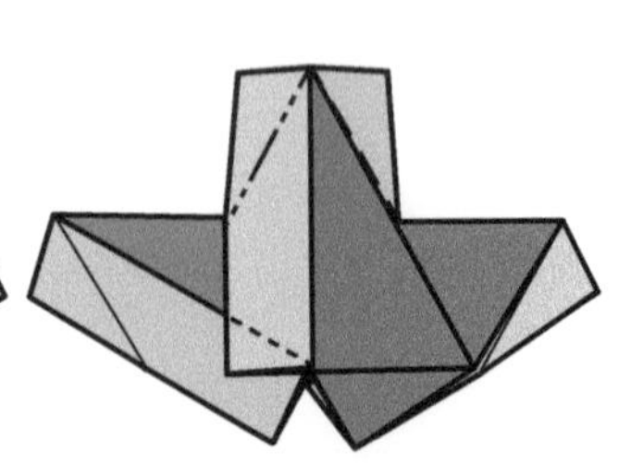

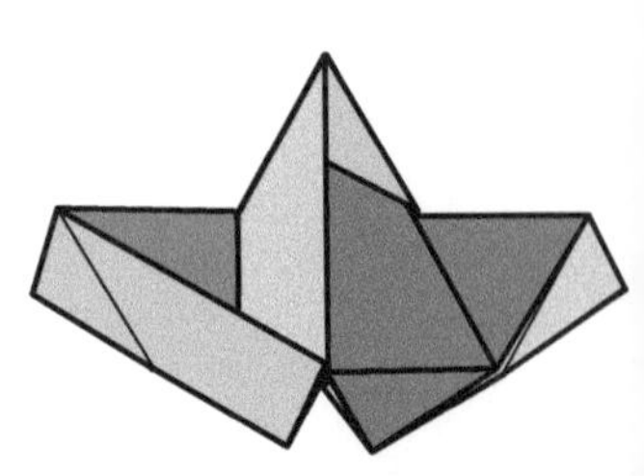

1 Interleave the flaps of both units. The first unit is at the bottom, and the second unit is at the top.

2 At the top, fold each tip into the pocket of the other unit. At the bottom, tuck each flap under the flap of the other unit.

3 Continue adding units in a clockwise direction.

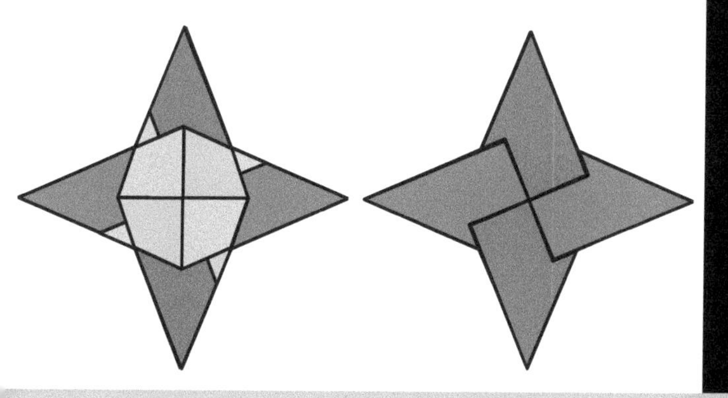

Star 4

Kite from Silver Rectangle

★

Use four silver rectangles at least A8

Finished model ratio 2.6

Step 3 gives this module a twist on the usual square from rectangle procedure. 80 gsm paper works fine, but stiff and strong paper may work better as the finished model only has a few layers of paper. In assembly step 4, can you find other ways to use the flaps of paper?

Module ★

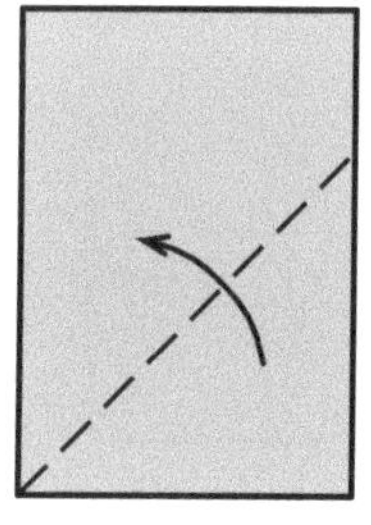

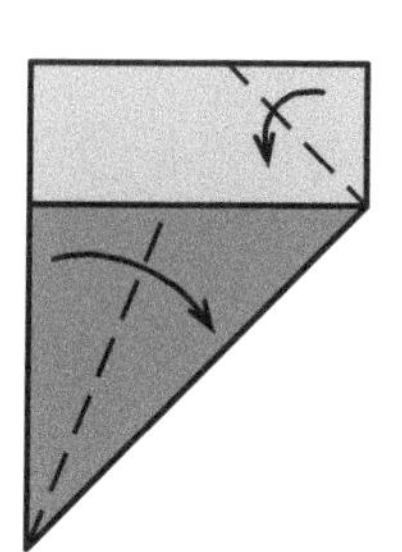

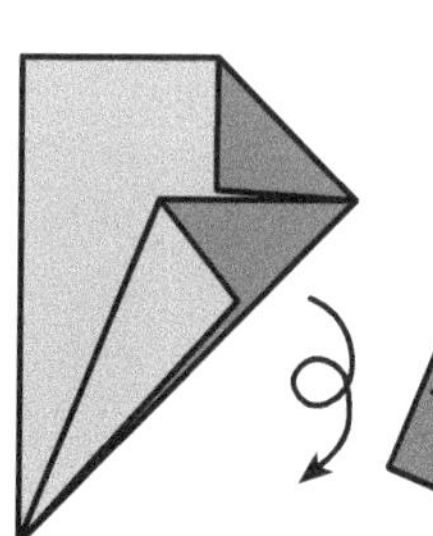

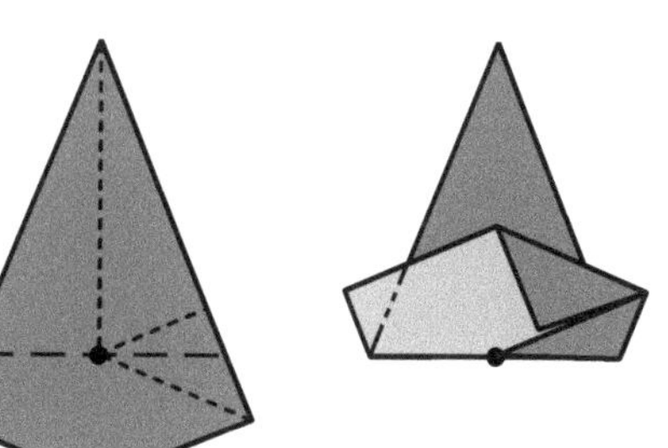

1 Bisect the lower left right angle.

2 Bisect the upper right right angle. Bisect the lower 45° angle. Turn over.

3 Fold the bottom flap up. The fold goes through the vertex underneath.

4 Mountain fold the little flap on the left and unfold.

Assembly ★

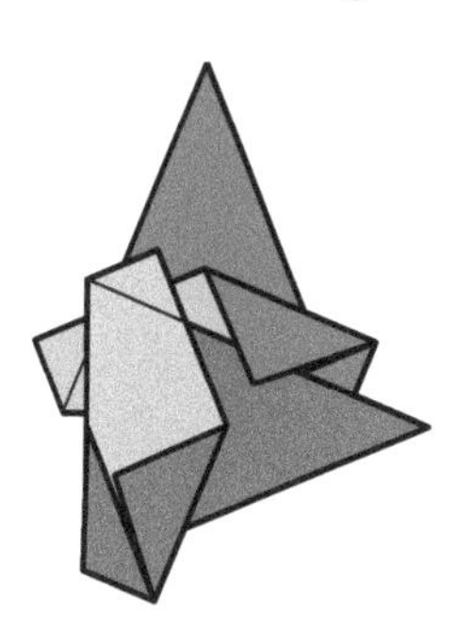

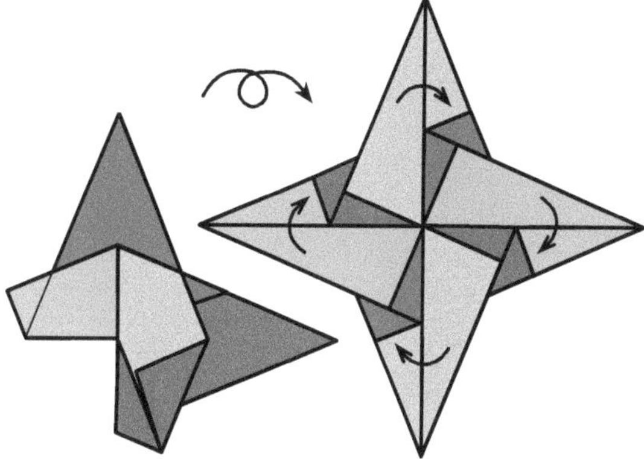

1 Hook the folded flap of the first unit onto the second unit and slide left.

2 Mountain fold the little flap of the second unit behind to lock.

3 Two units joined. Continue adding units in a clockwise direction.

4 When all units are joined, turn over and unfold the flaps.

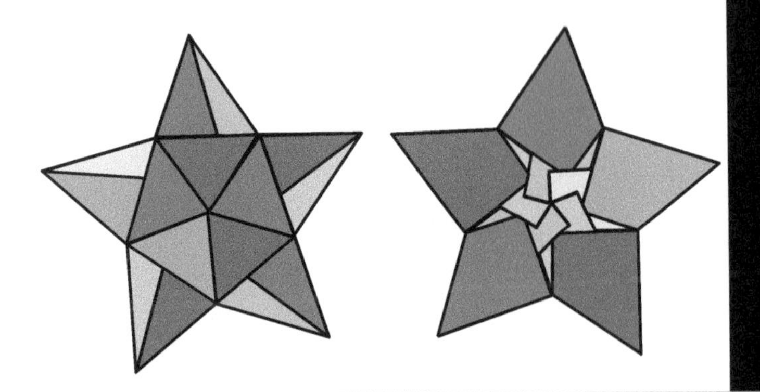

Star 5
from Half Silver Rectangle

★

Use five half silver rectangles, e.g. cut from A6
Finished model ratio 2.7

When cutting the silver rectangles, cut on the long mirror lines (as cutting on the short mirror lines would result in more silver rectangles). Although a little thick in the centre, it is acceptable. Even though not all flaps are secured, this star holds together well.

Module ★

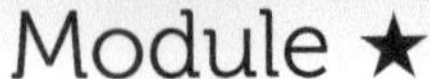

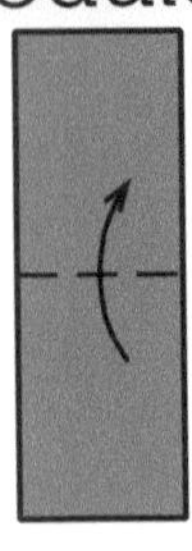

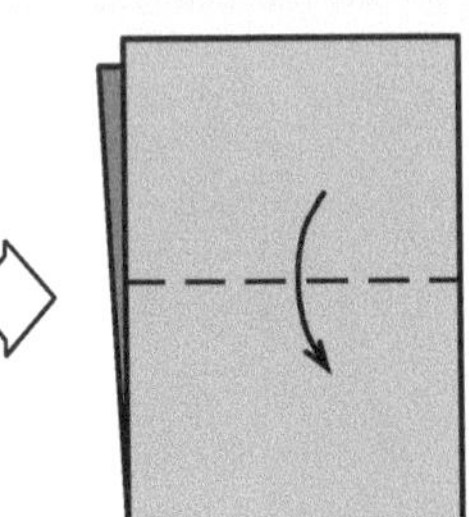

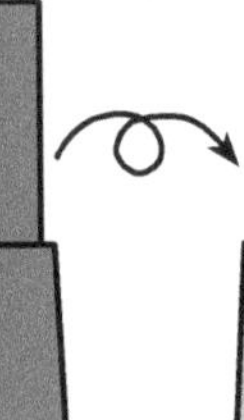

1 Fold the bottom edge to the top edge.

2 Fold the top layer in half.

3 Turn over.

4 Take the top right corner and fold the diagonal.

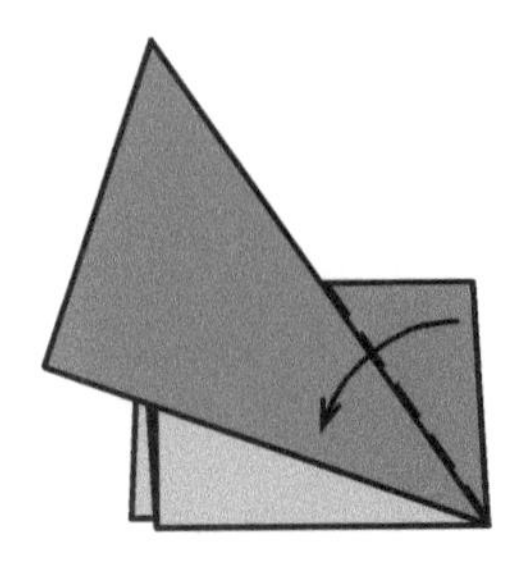

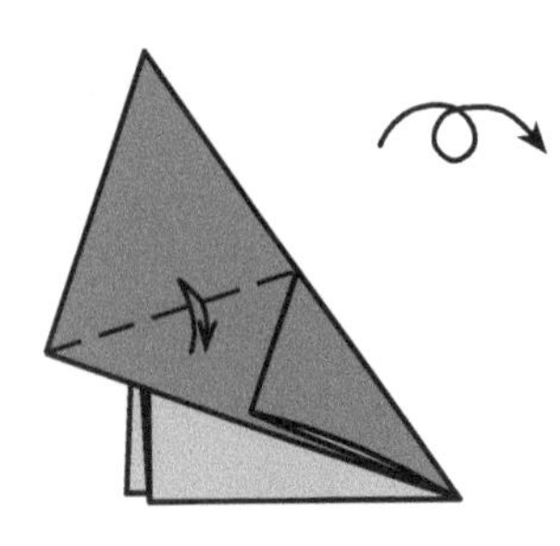

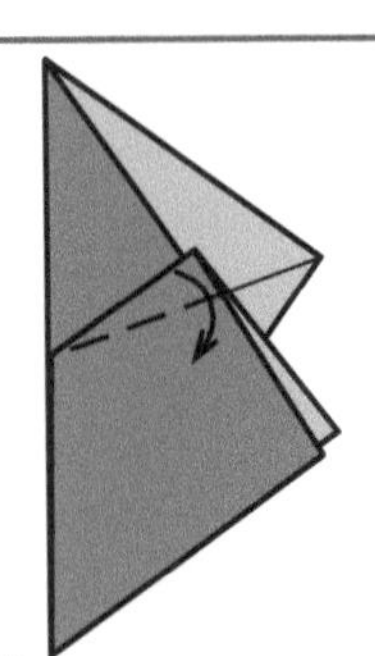

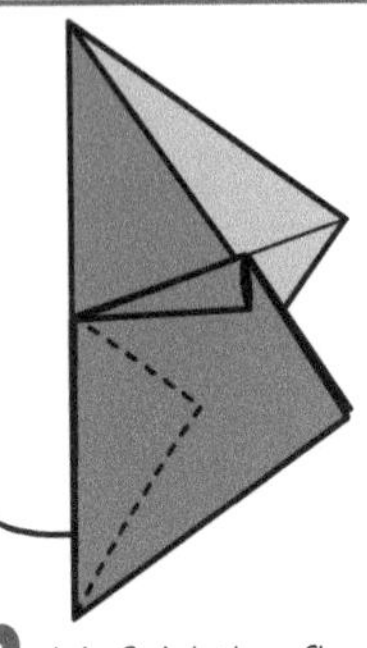

5 Fold the right flap along the diagonal.

6 Crease lightly and then turn over.

7 Fold using the crease from step 6.

8 Unfold the flap from behind.

Assembly ★

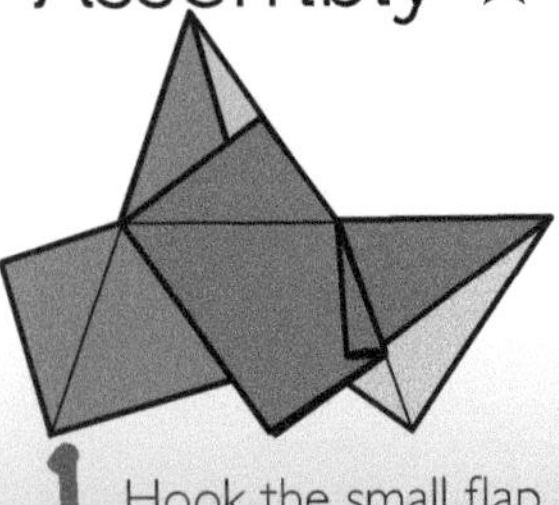

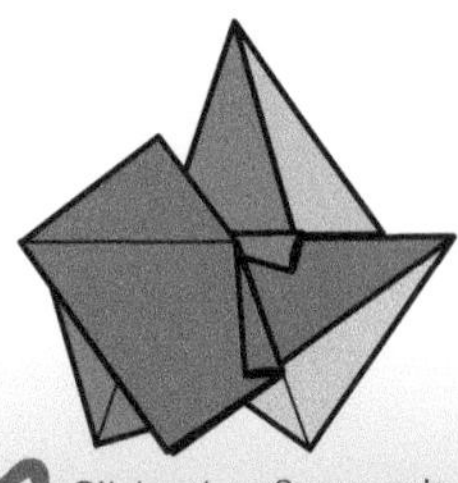

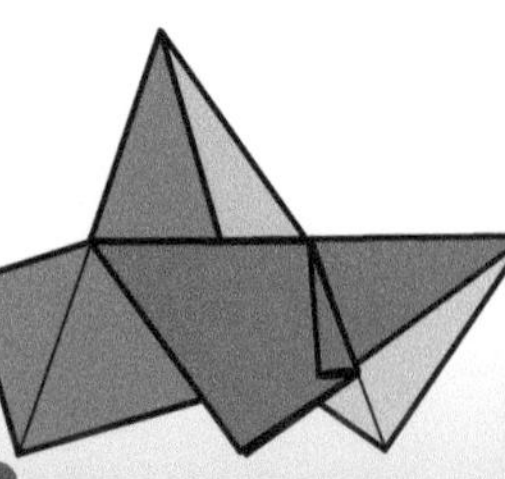

1 Hook the small flap of the first unit onto the second.

2 Slide the first unit left.

3 Mountain fold the flap of the second unit behind to lock.

4 Continue adding units in a clockwise direction.

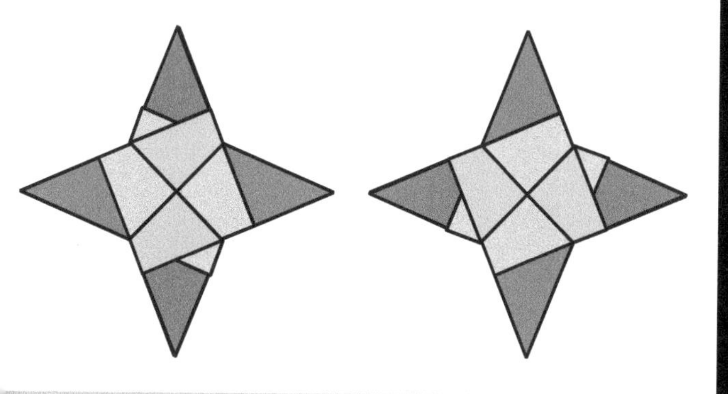

Star 4
Leftover Wrap

★
Use four leftover rectangles with short edges 40 mm or more
Finished model ratio 2.6

Try turning this model over: does it look identical, or do the colours change? You will find that it depends on the axis that you choose: the short axis preserves colours, but the long axis swaps colours.

Module ★

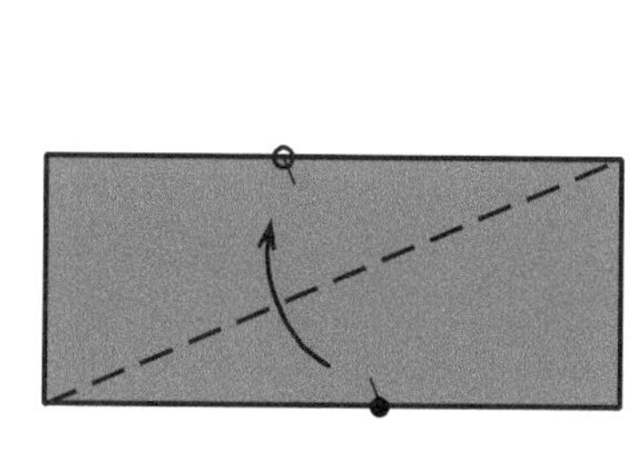

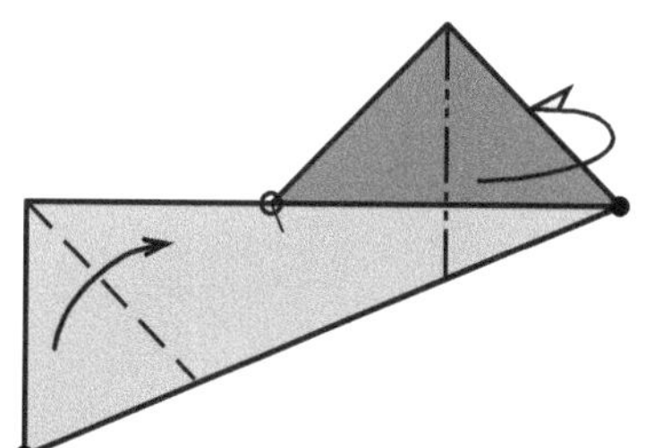

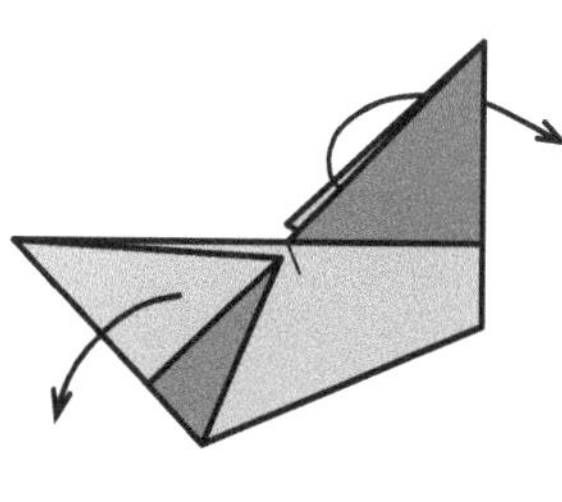

1 Join lower left and top right corners and pinch edges. Unfold and then join pinch marks to fold the diagonal.

2 Fold the left corner to the intersection of edges. Turn over and repeat.

3 Unfold the flaps.

Assembly ★

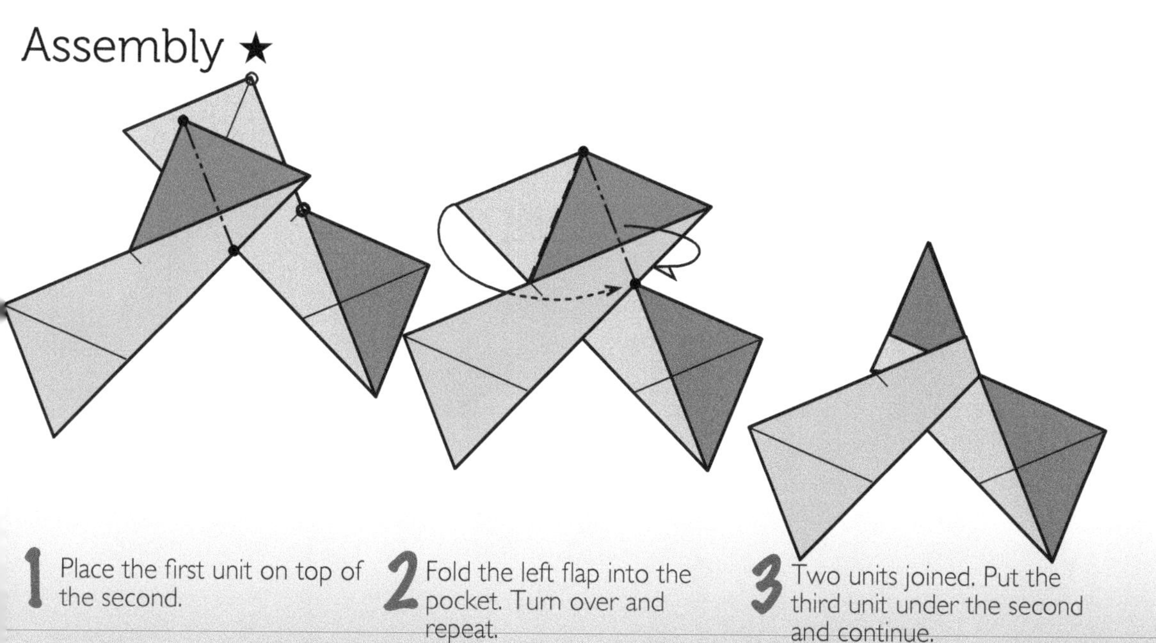

1 Place the first unit on top of the second.

2 Fold the left flap into the pocket. Turn over and repeat.

3 Two units joined. Put the third unit under the second and continue.

Star 6
Bronze Wrap

★

Use six bronze rectangles with short edges 40 mm or more
Finished model ratio 2.0

This model is similar to the previous model. In step1, you can join opposite corners together and fold all the way through as the crease will not show in the final model. However, pinching only where needed is better for the integrity of the paper.

Module ★

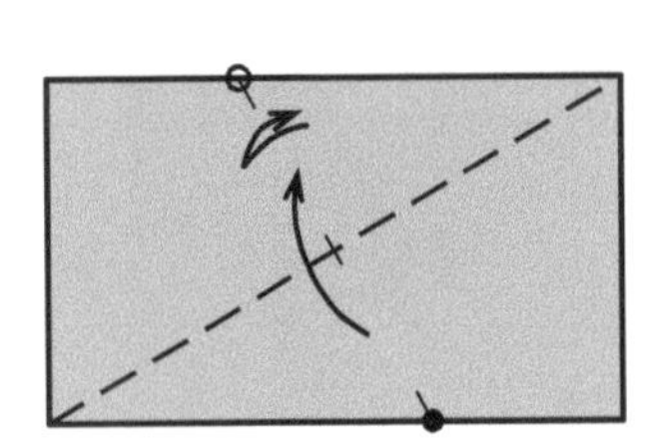
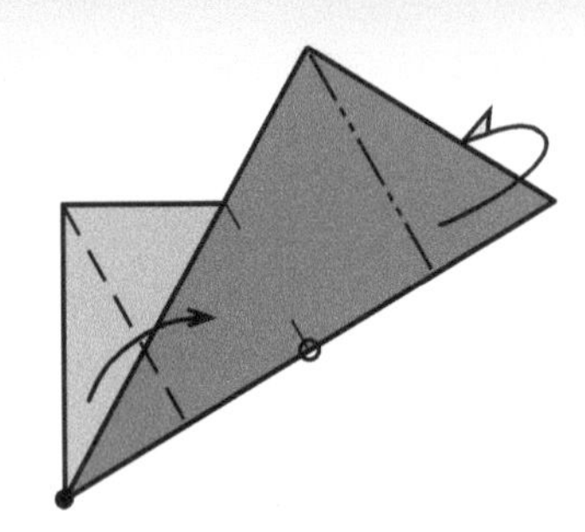
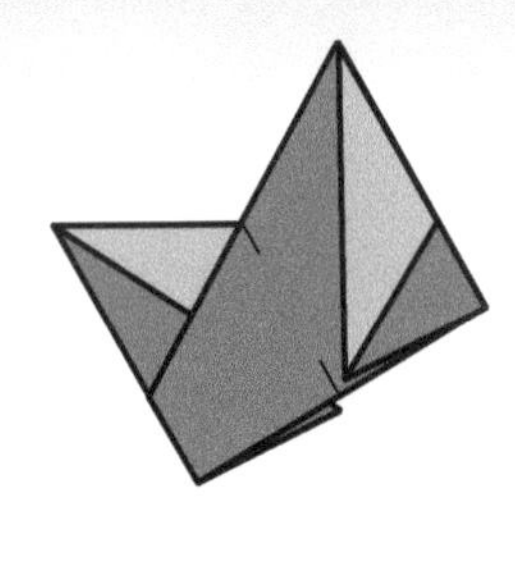

1 Join lower left and top right corners and pinch edges and centre. Join the pinch marks to fold the diagonal.

2 Fold the left corner to the centre. Turn over and repeat.

3 Unfold the flaps.

Assembly ★

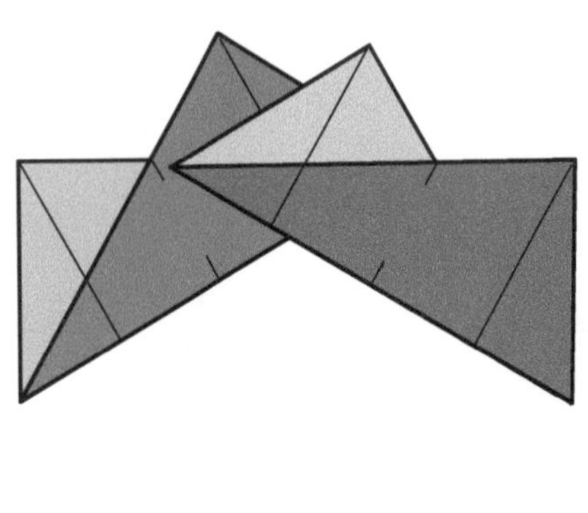
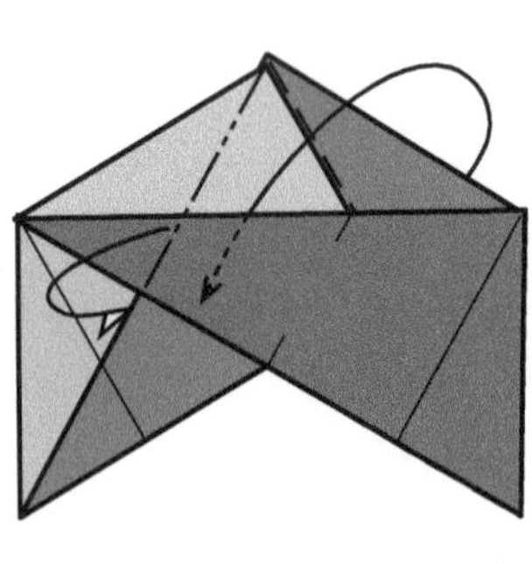
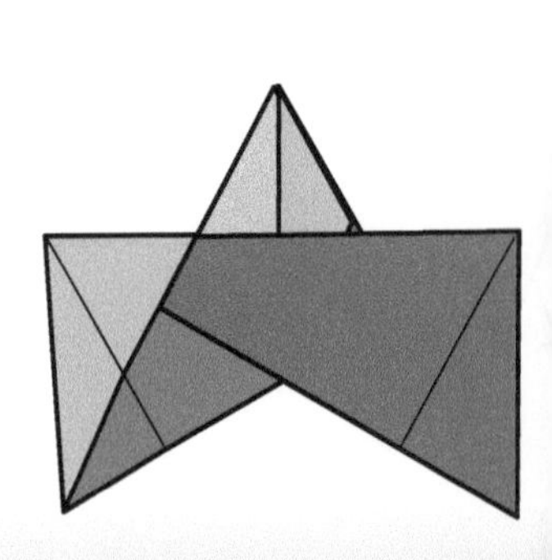

1 Place the second unit on top of the first.

2 Fold the right flap into the pocket. Turn over and repeat.

3 Two units joined. Put the third unit on top of the second and continue.

Star 8
Leftover
Wrap 45°

★
Use eight leftover rectangles with short edges 50 mm or more
Finished model ratio 2.7

The units are stable only when all are assembled. Even so, the finished model will disassemble if the points are pulled apart too far, or spun too much.

Module ★

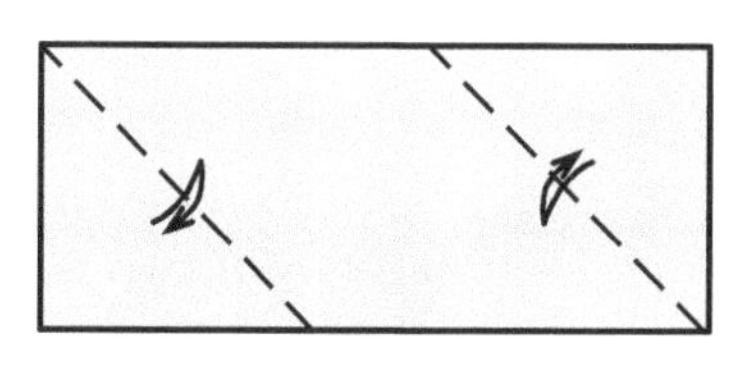

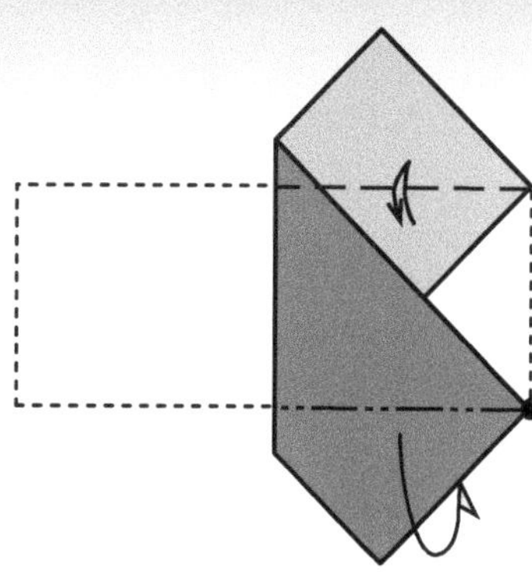

1 Fold the diagonal of the square on the left. Rotate 180° and repeat. Make at least two, the first being a template.

2 Put a unit on top of the template. Firmly hold in place and then fold the lower left corner to the right.

3 Remove the template. Fold on the existing creases.

Assembly ★★

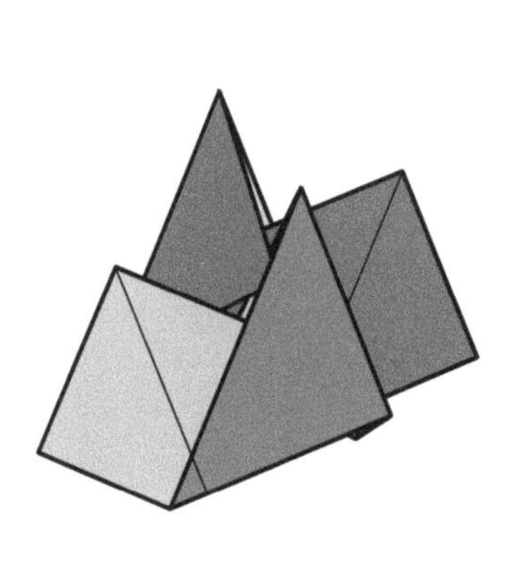

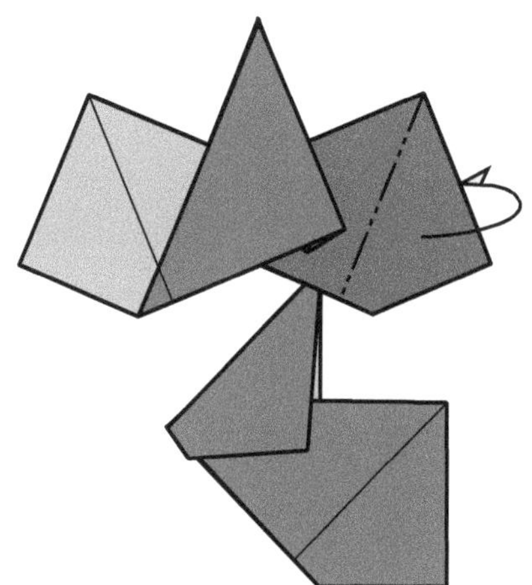

1 Interleave the rear flaps of the first unit and the front flap of the second unit.

2 Slide together.

3 Mountain fold the right flap of the second unit. Then put third unit under second unit and join.

Star 8
Leftover Wrap 22.5°

★★

Use eight leftover rectangles with short edges 50 mm or more
Finished model ratio 2.7

Unlike the previous models, this model changes colour when turned on the short axis instead of the long axis.

Module ★★

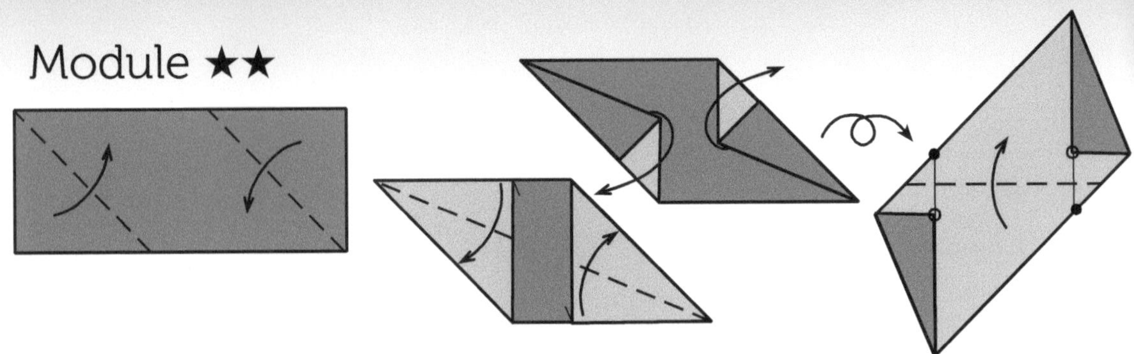

1 Fold the diagonal of the square on the left. Rotate 180° and repeat.

2 Bisect the 45° flaps.

3 Unfold and turn over.

4 Fold the ends of the diagonals onto the original corners of the rectangle.

Assembly ★★

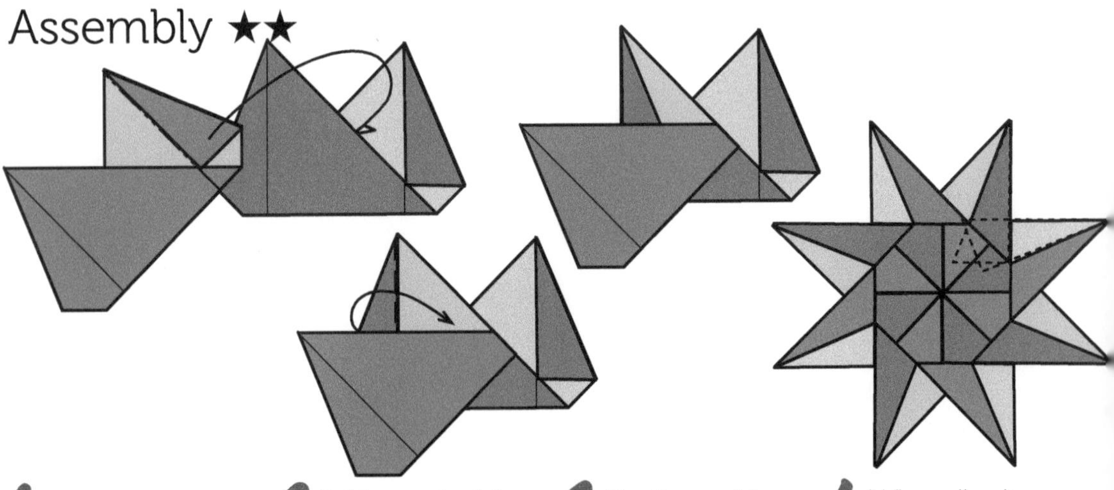

1 Mountain fold the right flap of the first unit and hook into the second.

2 Fold over the left edge of the second unit.

3 Continue adding units in an anti-clockwise direction.

4 When all units are assembled, interlock the flaps.

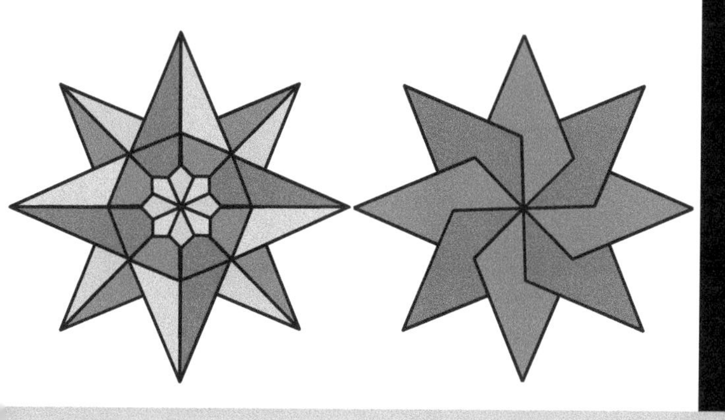

Medallion Star

★★★
Use eight 150 mm squares
Finished model ratio 1.5

Each unit has up to six layers of paper and the central octagon of the star has twelve layers. Therefore this star works best with thinner paper, e.g. 15 cm origami paper, or larger paper.

Module ★★

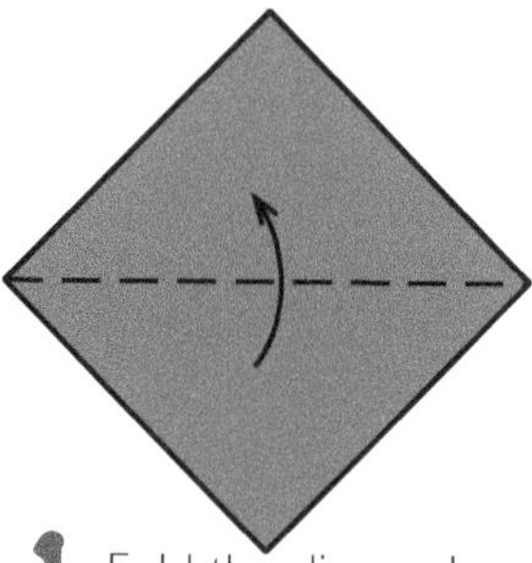
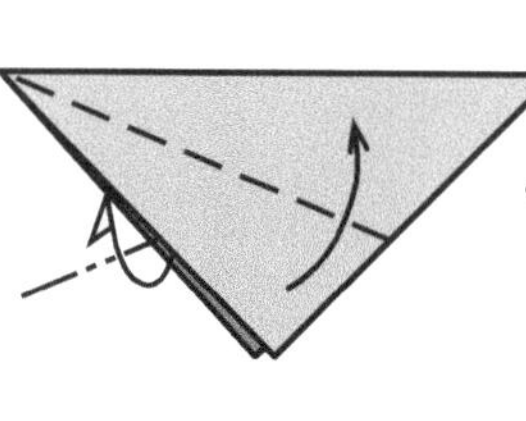
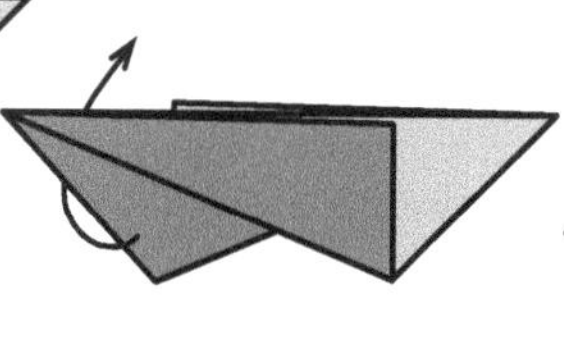
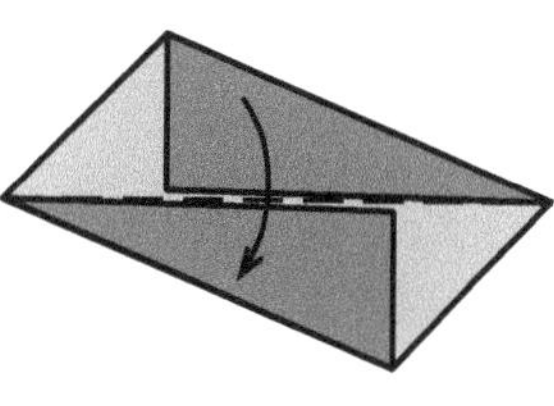

1 Fold the diagonal.

2 Bisect the left 45° angle. Turn over and repeat.

3 Unfold the diagonal from behind.

4 Fold on the diagonal with the flaps inside.

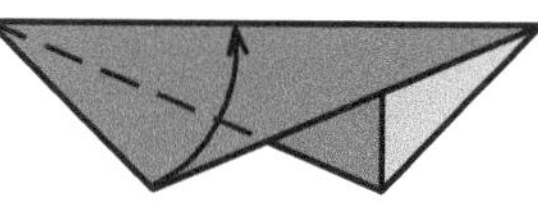
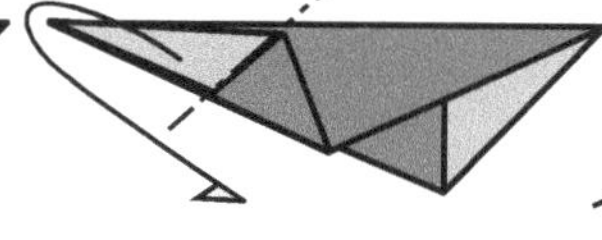
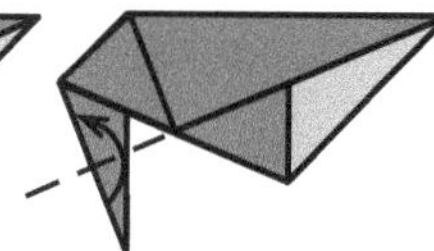
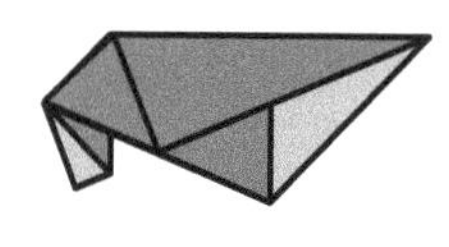

5 Bisect the left 45° angle.

6 Use the raw edge to guide the mountain fold of the flap behind.

7 Fold the tip to the nearest obtuse angle.

8 Unfold and return to step 6.

Assembly ★★★

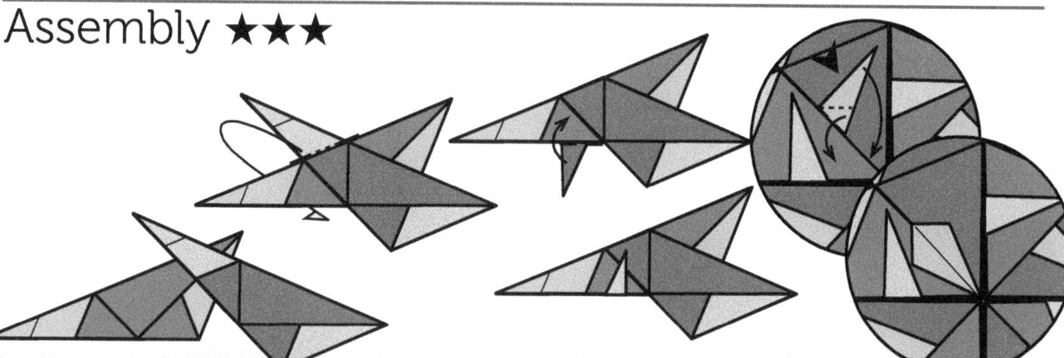

1 Put the second unit on top of the first.

2 Mountain fold the flap of the second unit behind.

3 Fold the tip of the second unit over the first unit.

4 When all units are joined, open and squash the tips to lock.

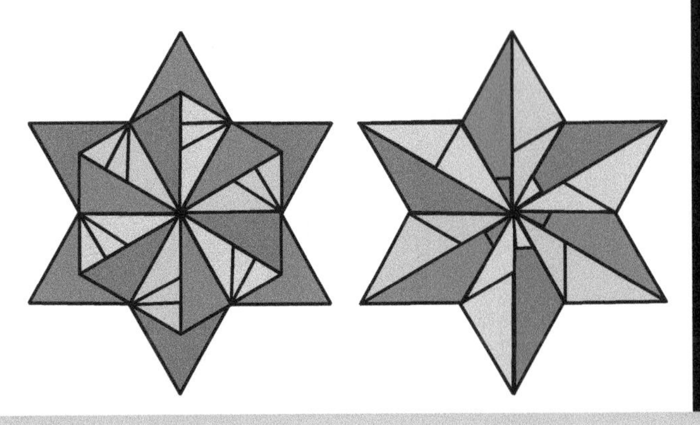

Star 6 Fancy Wrap

★★★

Use six 75 mm squares or larger
Finished model ratio 1.7

In step 4, the right crease is not exact: it uses the approximation $\sqrt{3}\approx 7\div 4$, or $1.73\approx 1.75$. If joining the last unit is too hard, try sliding two groups of three units together.

Module ★★★

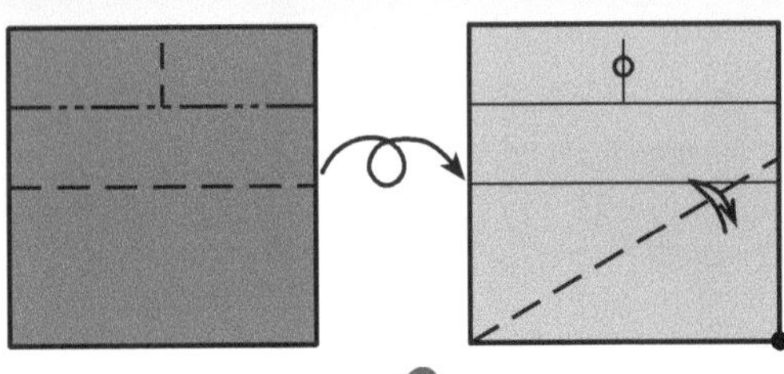

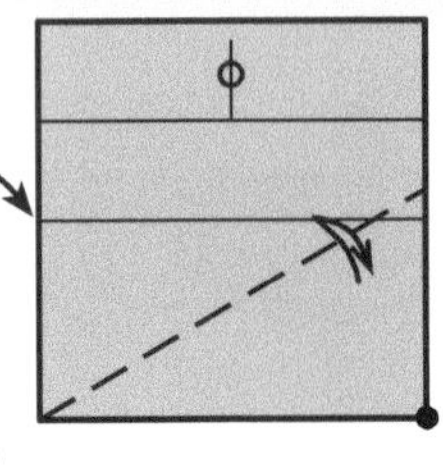

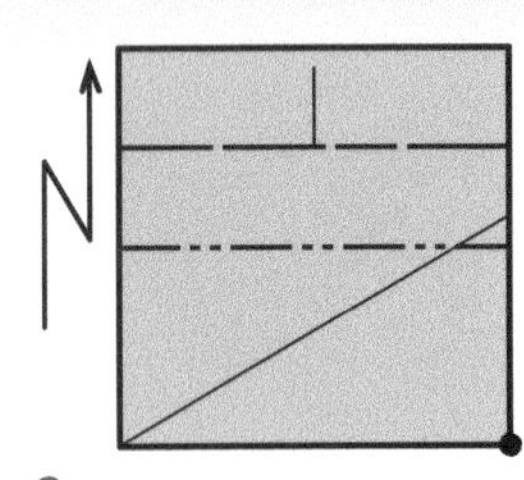

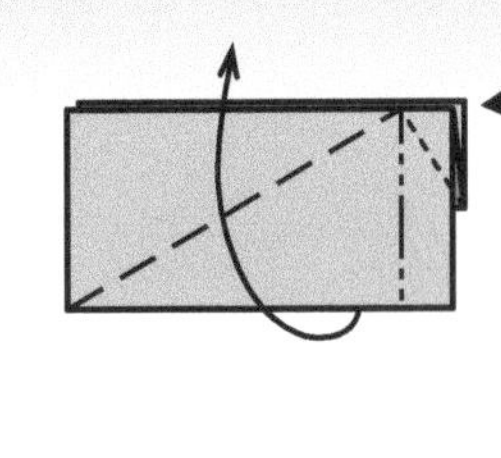

1 Precrease and turn over.

2 Crease a 60° fold.

3 Pleat using existing creases.

4 Lift corner and squash.

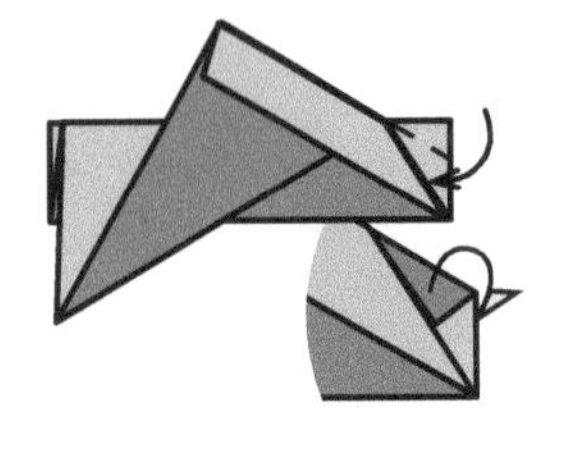

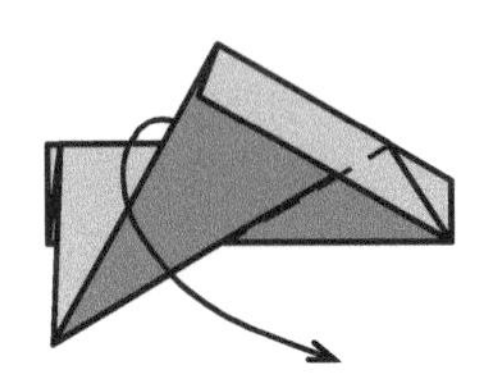

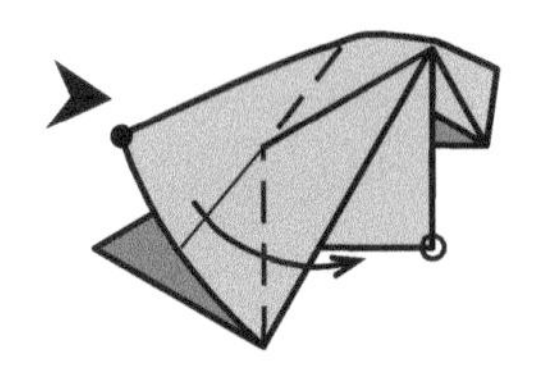

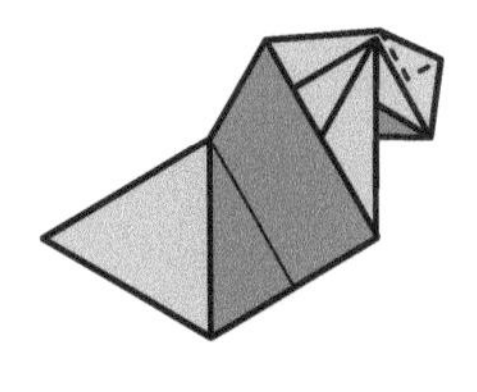

5 Fold the little flap. Unfold and fold behind.

6 Fold all layers except the furthest layer.

7 Squash the flap.

8 Fold flap down.

Assembly ★★★

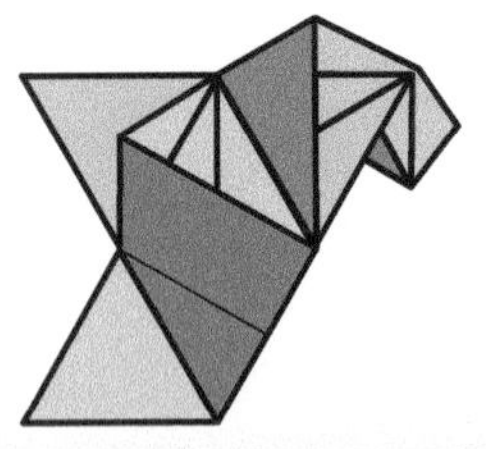

1 Hook the second unit onto the first.

2 Slide right.

3 Continue adding in an anticlockwise direction.

4 The last unit can be hard to join.

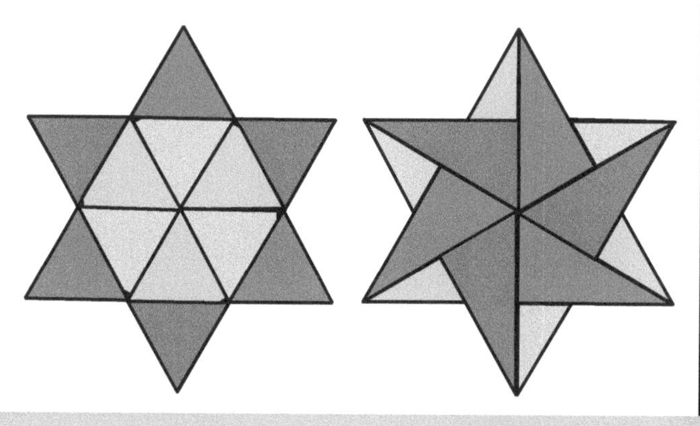

Star 6
Equilateral Triangle from Rectangle

★

Use six silver rectangles, e.g. A7
Finished model ratio 2.0

To avoid the small gaps, use a rectangle 1:5/(2√3), i.e. 1:1.443. You can use longer rectangles, e.g. a bronze rectangle, which needs to have one sixth folded back.

Module ★

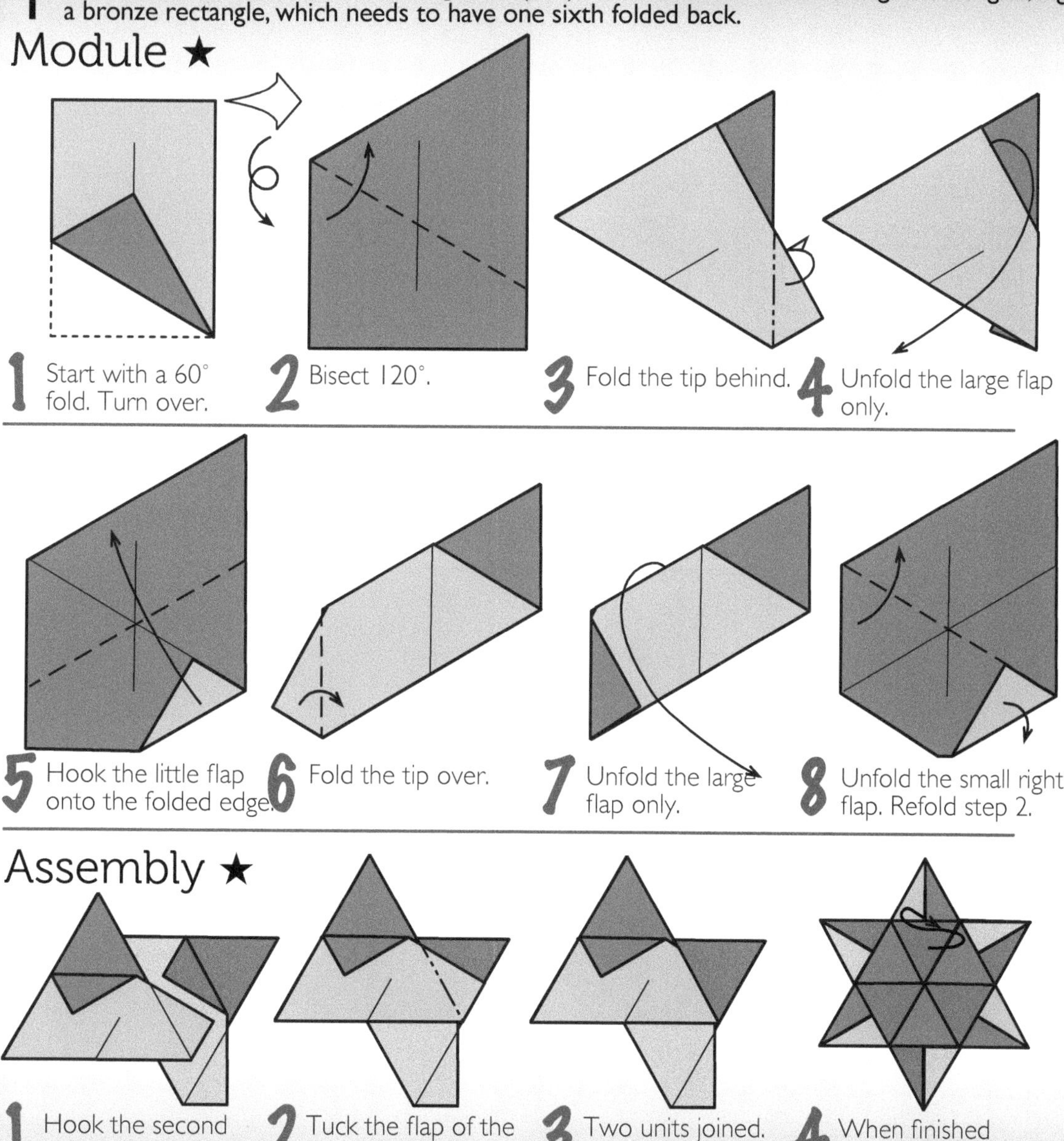

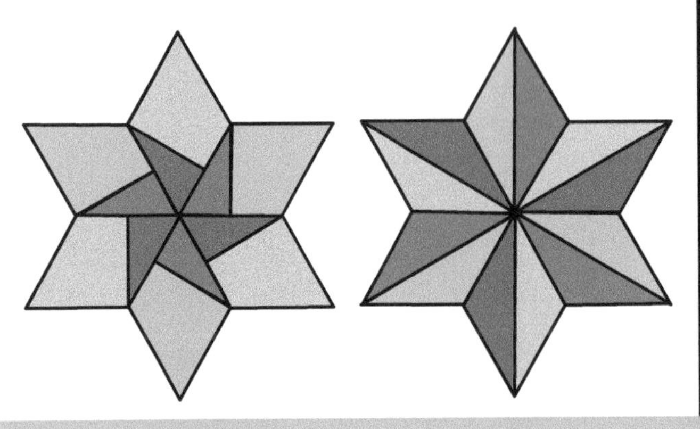

Star 6
Bronze Triangle Wheel

★★

Use six bronze rectangles
Finished model ratio 2.0

The layers can become thick, so you could cut off the flaps to start with a diamond. Alternatively, use thin or large paper. There are two other models that are related to this model: can you find them?

Module ★★

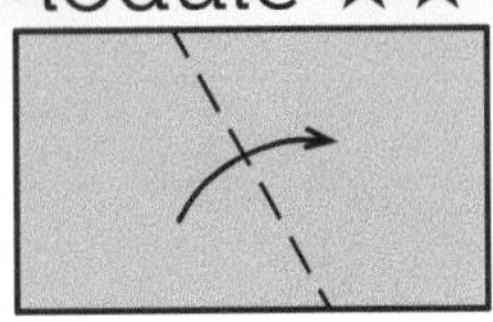

1 Join opposite corners together.

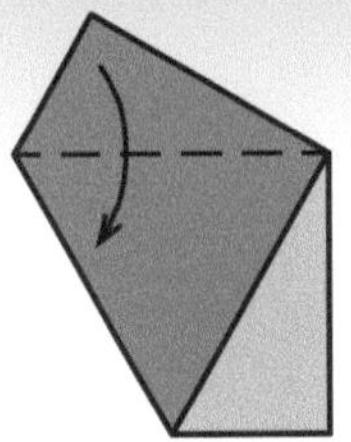

2 Bisect 120°.

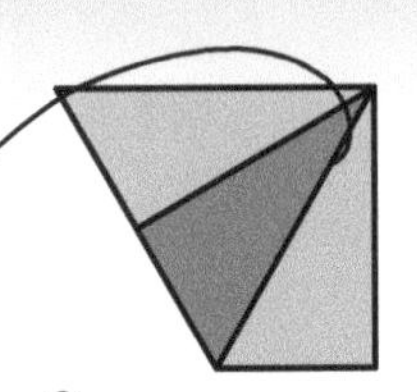

3 Unfold the first fold only.

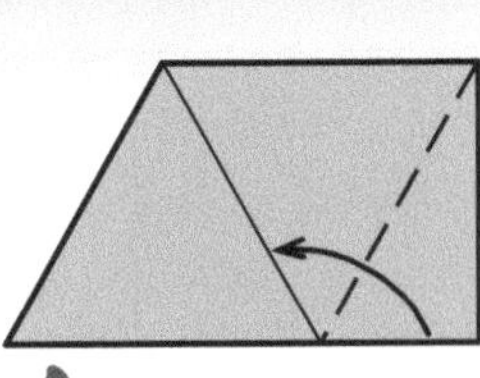

4 Bisect 120°.

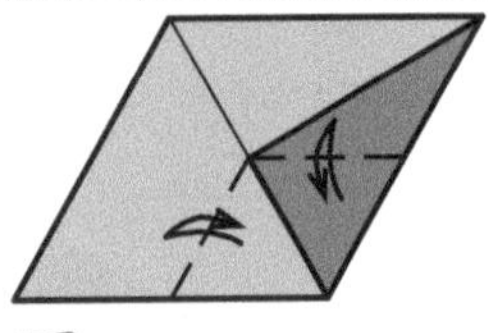

5 Precrease.

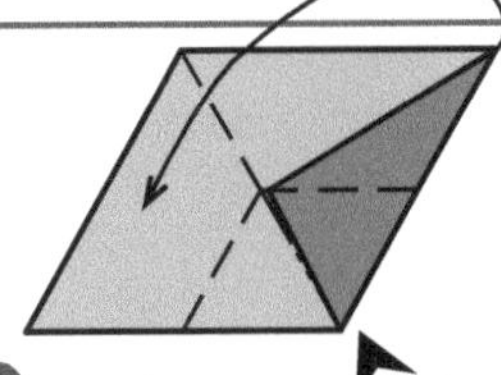

6 Inside reverse fold.

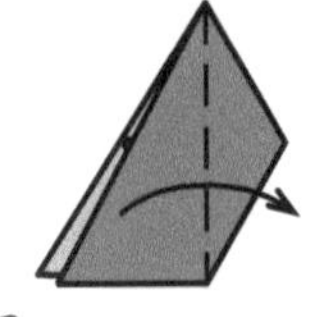

7 Fold the flap right.

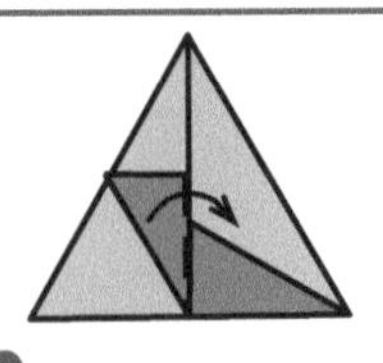

8 Fold the flap right.

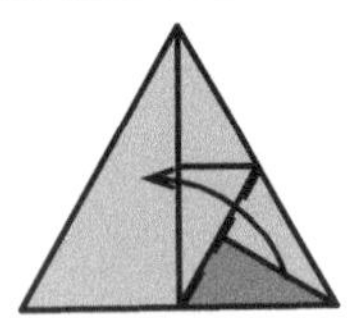

9 Fold the flap over the folded edge.

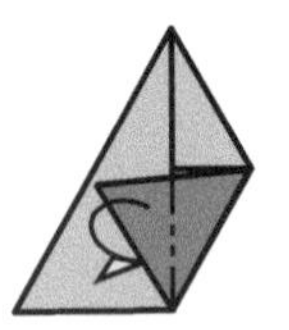

10 Fold inside.

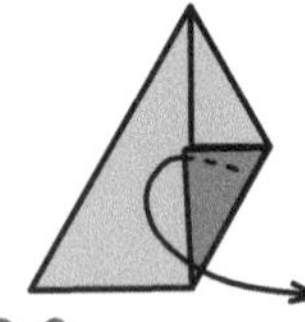

11 Pull out and return to step 8.

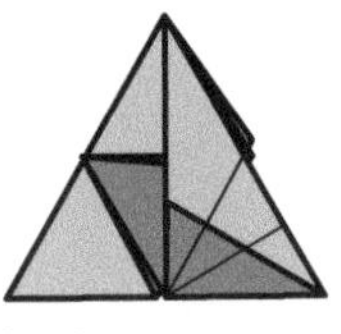

12 Unit complete.

Assembly ★★

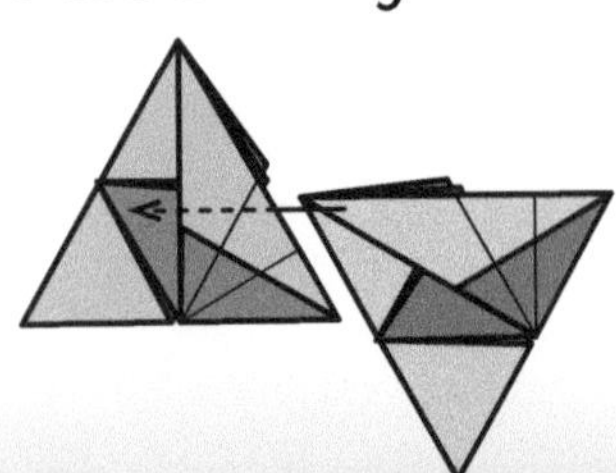

1 Insert the second unit into the reverse fold of the first.

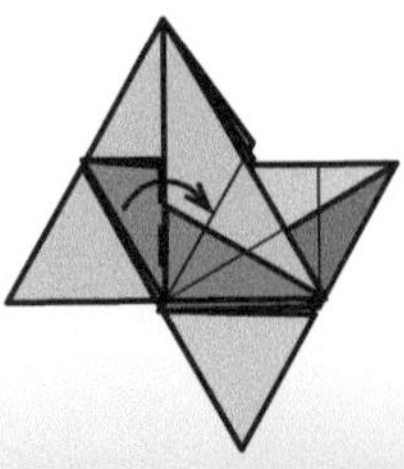

2 Repeat module steps 8 to 11. Fold over.

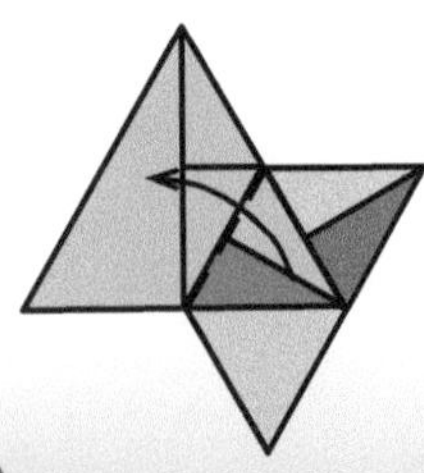

3 Fold over.

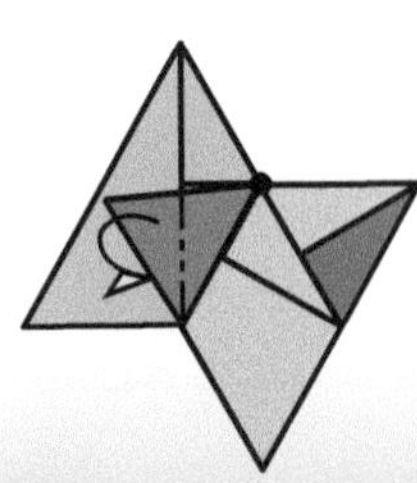

4 Tuck inside. The dot shows the centre of the star.

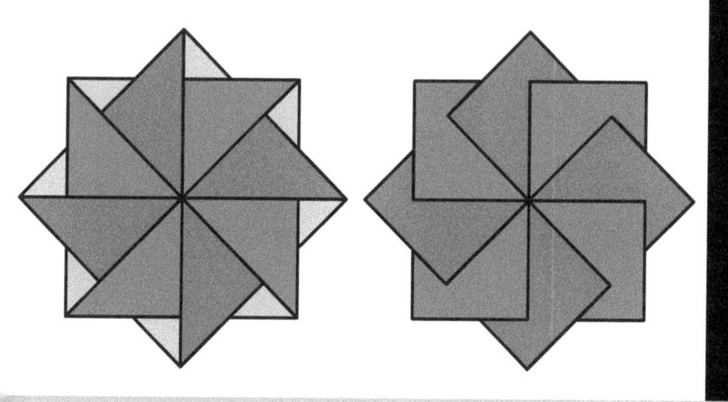

Intersecting Squares

★★★
Use eight 75 mm squares or larger
Finished model ratio 1.4

Using paper that is the same colour on both sides will enhance the illusion that two squares are intersecting each other.

Module ★★

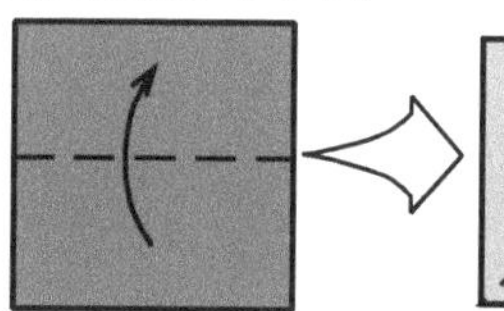

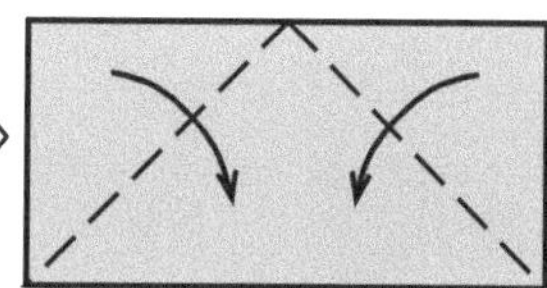

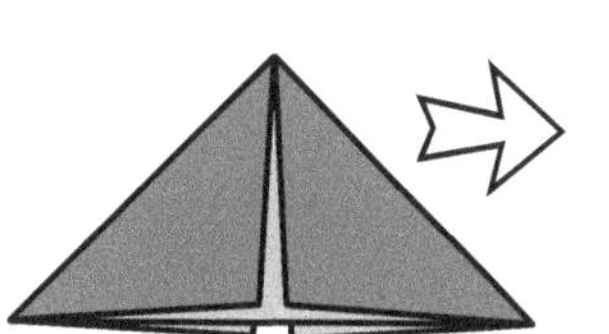

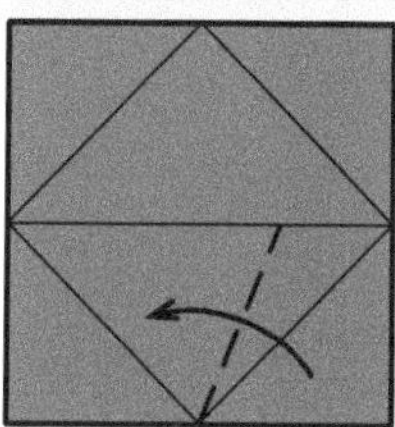

1 Fold the bottom edge to the top edge.

2 Bisect the lower left and right right angles by folding the left and right edges to the bottom edge.

3 Open up.

4 Only fold the lower half of the square: bisect 135° by folding the right edge to the left crease line.

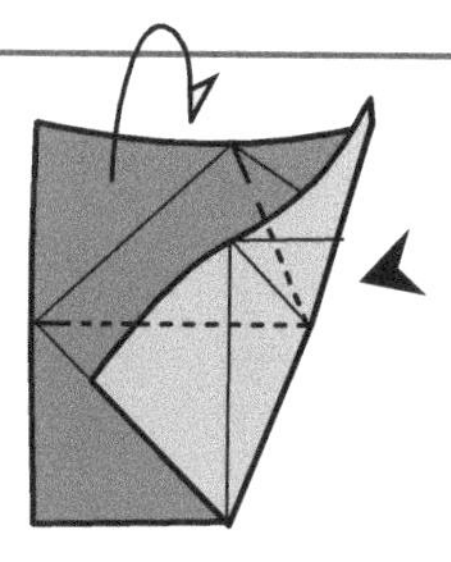

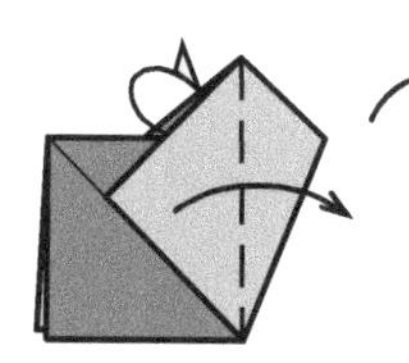

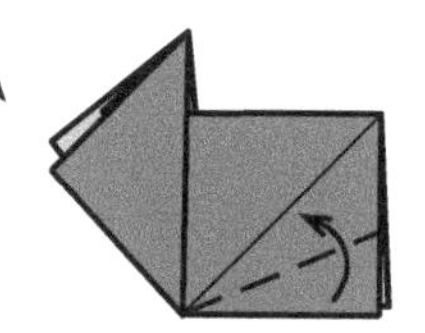

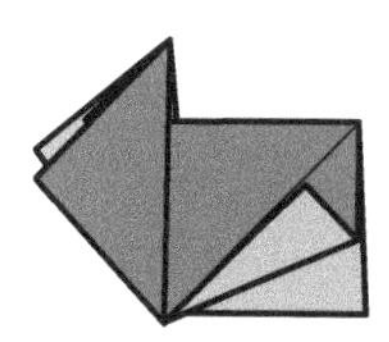

5 Mountain fold in half, squashing the right side.

6 Fold the flap to the right. Turn over and fold the flap to the left.

7 Only fold the front flap.

8 Unit complete.

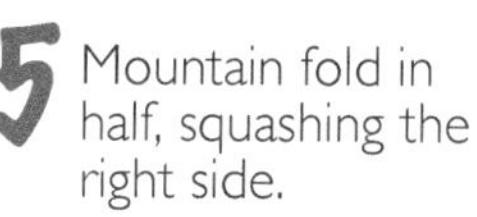

Assembly ★★

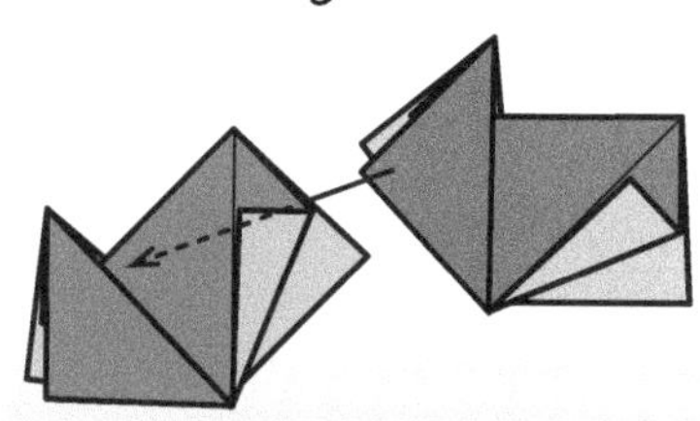

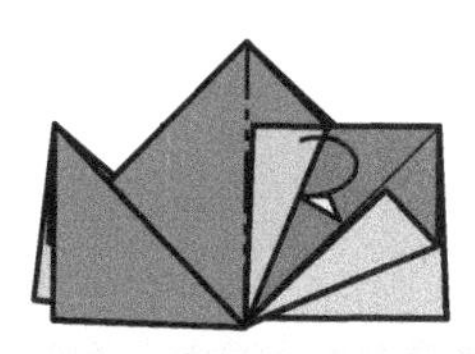

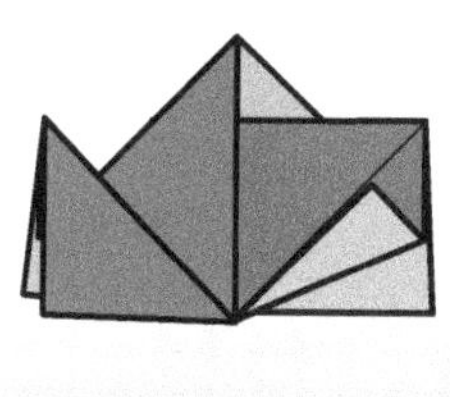

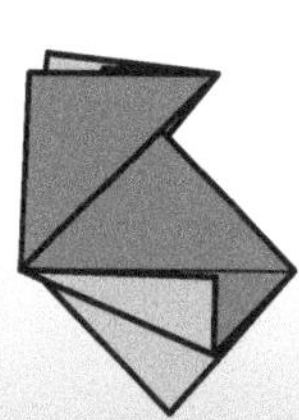

1 Insert both flaps of the second unit into the pocket of the first unit.

2 On the front only, mountain fold inside to lock. Add more units in a clockwise direction.

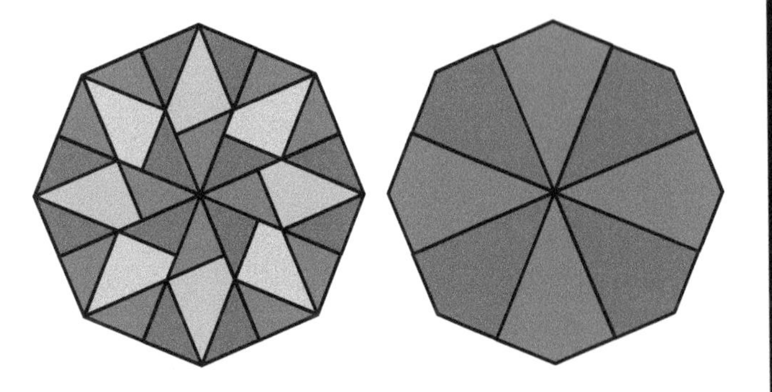

Quilt Star

★★

Use eight 75 mm squares with different colours on each side
Finished model ratio 1.5

Some patchwork quilts use isosceles right-angled triangles to make patterns. To highlight the eight-pointed star inside, use paper that is white on one side and one of two colours on the other.

Module ★★

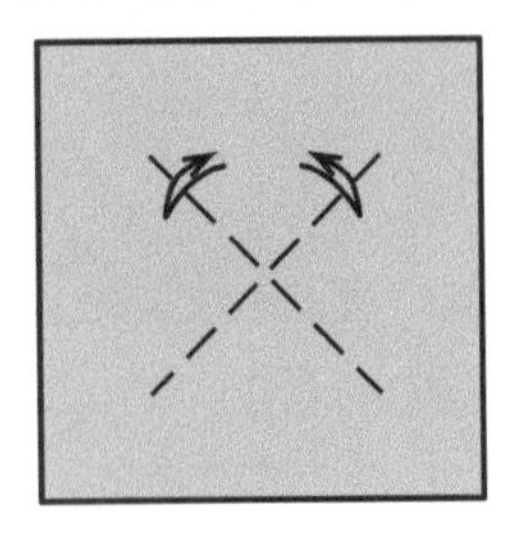

1 Crease partial diagonals.

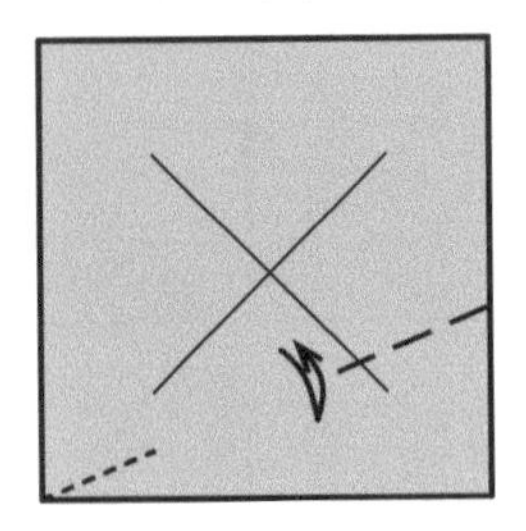

2 Bisect the lower left 45° angle. Only fold from the diagonal to the right edge.

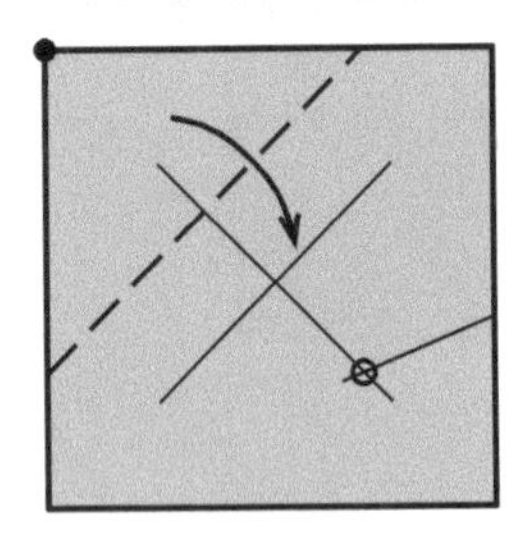

3 Fold the top left corner to the intersection of creases.

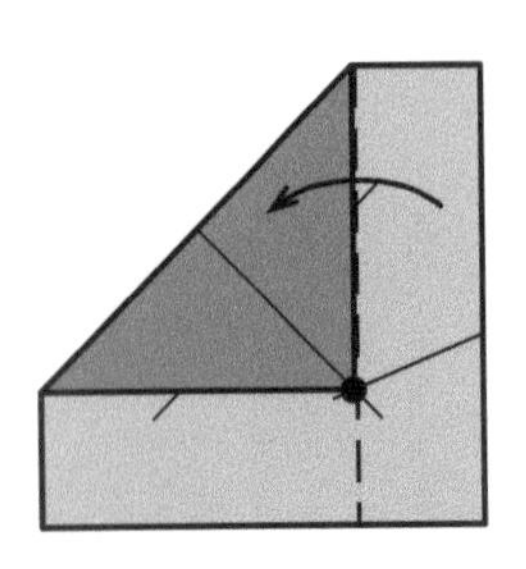

4 Fold the right edge over the the flap.

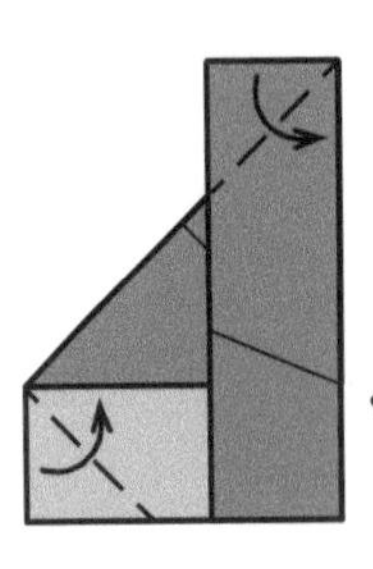

5 Fold the top and left corners and then unfold.

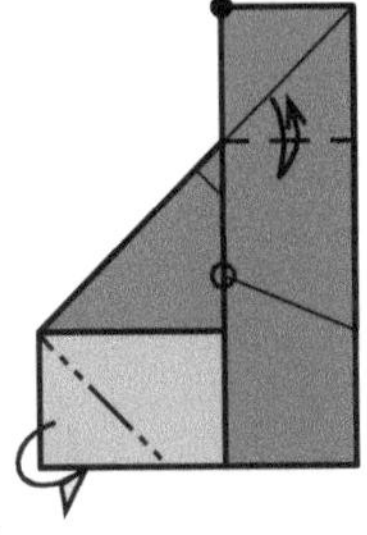

6 Fold the top down and unfold. Mountain fold the left corner.

Assembly ★★

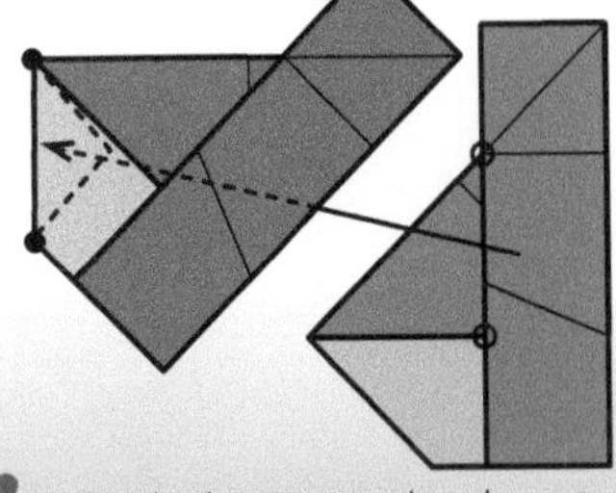

1 Hook the second unit onto the first unit.

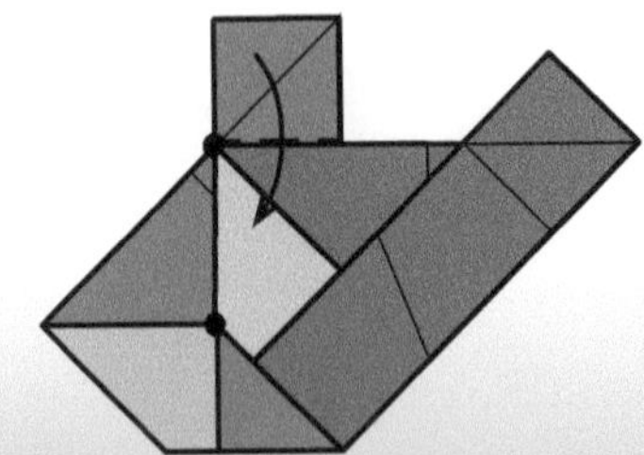

2 Fold down.

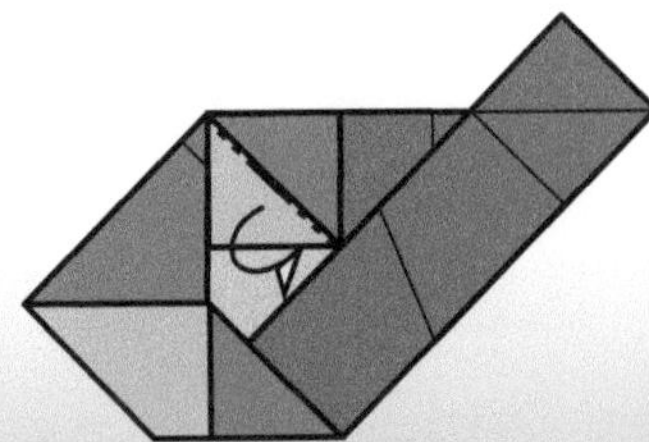

3 Fold the small flap behind into the second unit to lock.

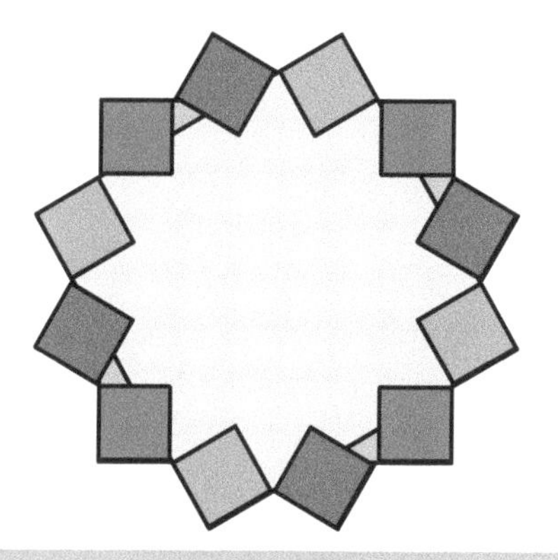

Three Four Dodecagon Squares

★★

Use three 15 cm squares with different colours on each side
Finished model ratio 1.0

This model works best from paper that is white on one side and coloured on the other. It was inspired by a two-piece octagram by Wayne Brown and a two-piece hexagram by Lewis Simon [Jackson 82, p. 19-21]. Try making smaller models and stacking them for diminishing squares. I later found out that Wayne had created a this 12-pointed star and 16-pointed star from four squares.

Module ★★

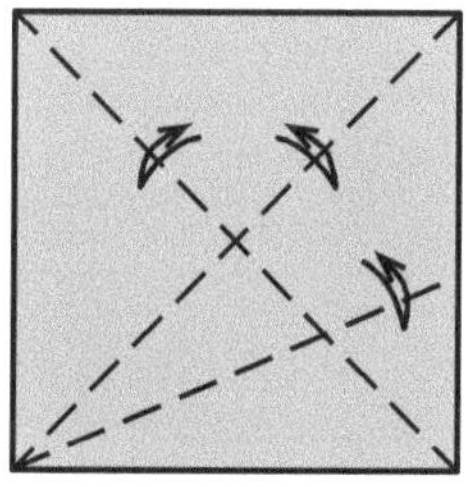

1 Crease the diagonals. Bisect the lower left 45° angle. Rotate 90°.

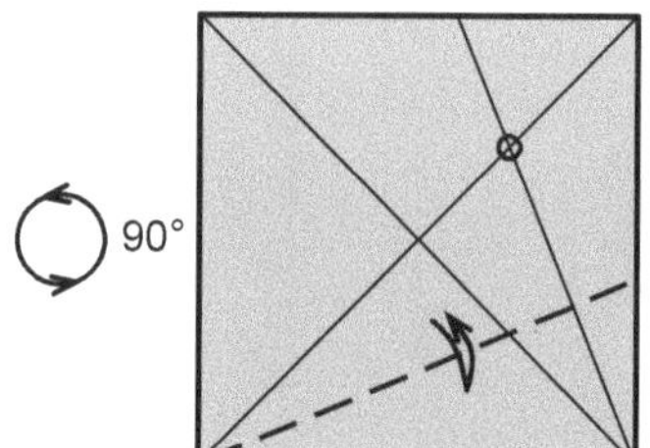

2 Bisect the lower left 45° angle. Rotate 90°. Repeat twice more.

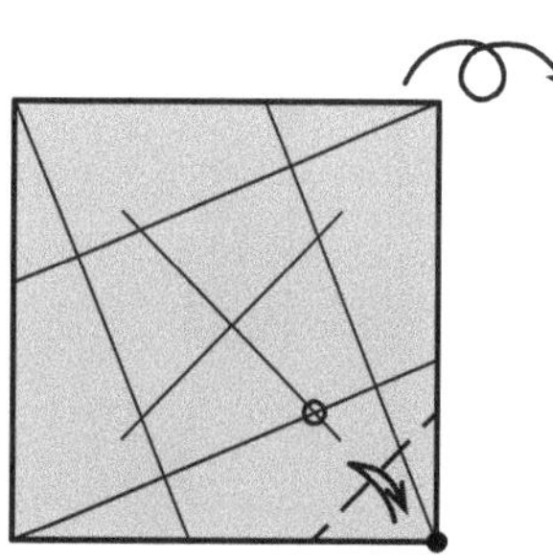

3 Crease the bottom right corner to the intersection of creases. Repeat three times. Turn over.

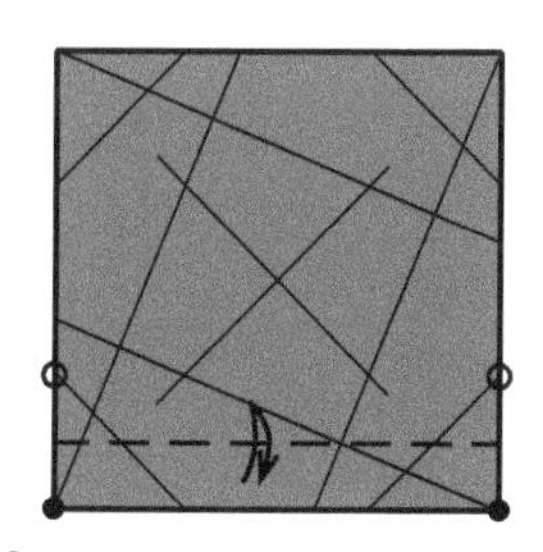

4 Fold the edge and unfold. Repeat three times.

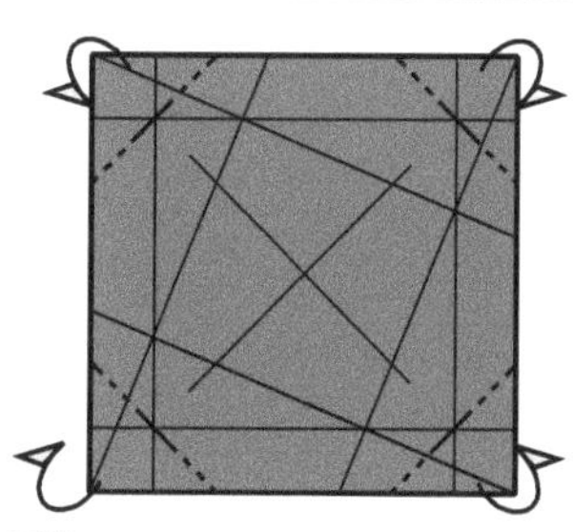

5 Fold the corners behind.

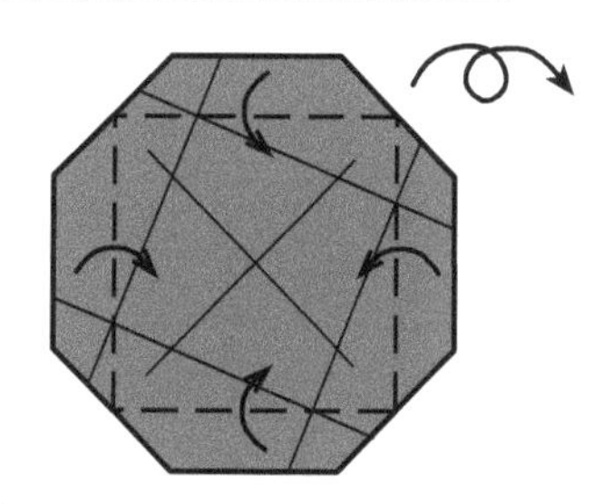

6 Fold the edges in. Turn over. Make two more units.

Assembly ★★

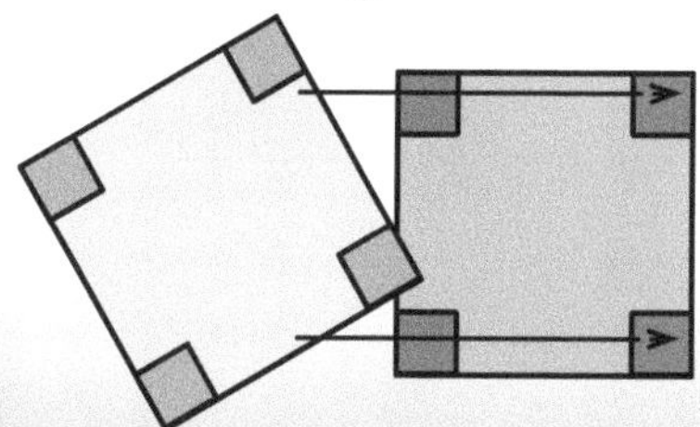

1 Insert the second unit into the first.

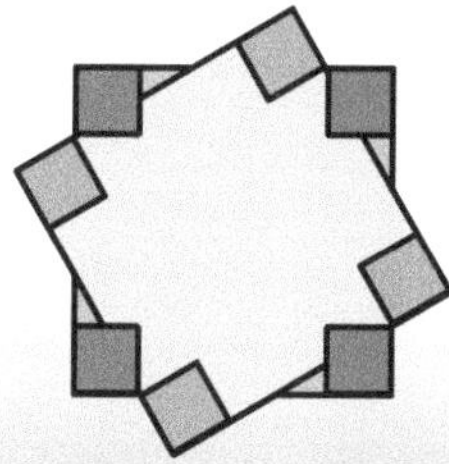

2 Insert these two units into the third unit.

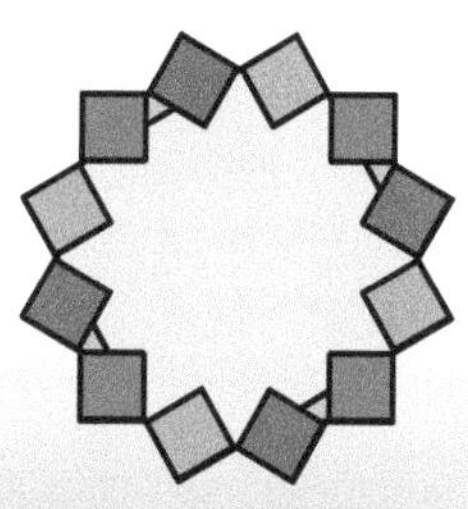

3 Three units joined.

Chapter 10

Sliders

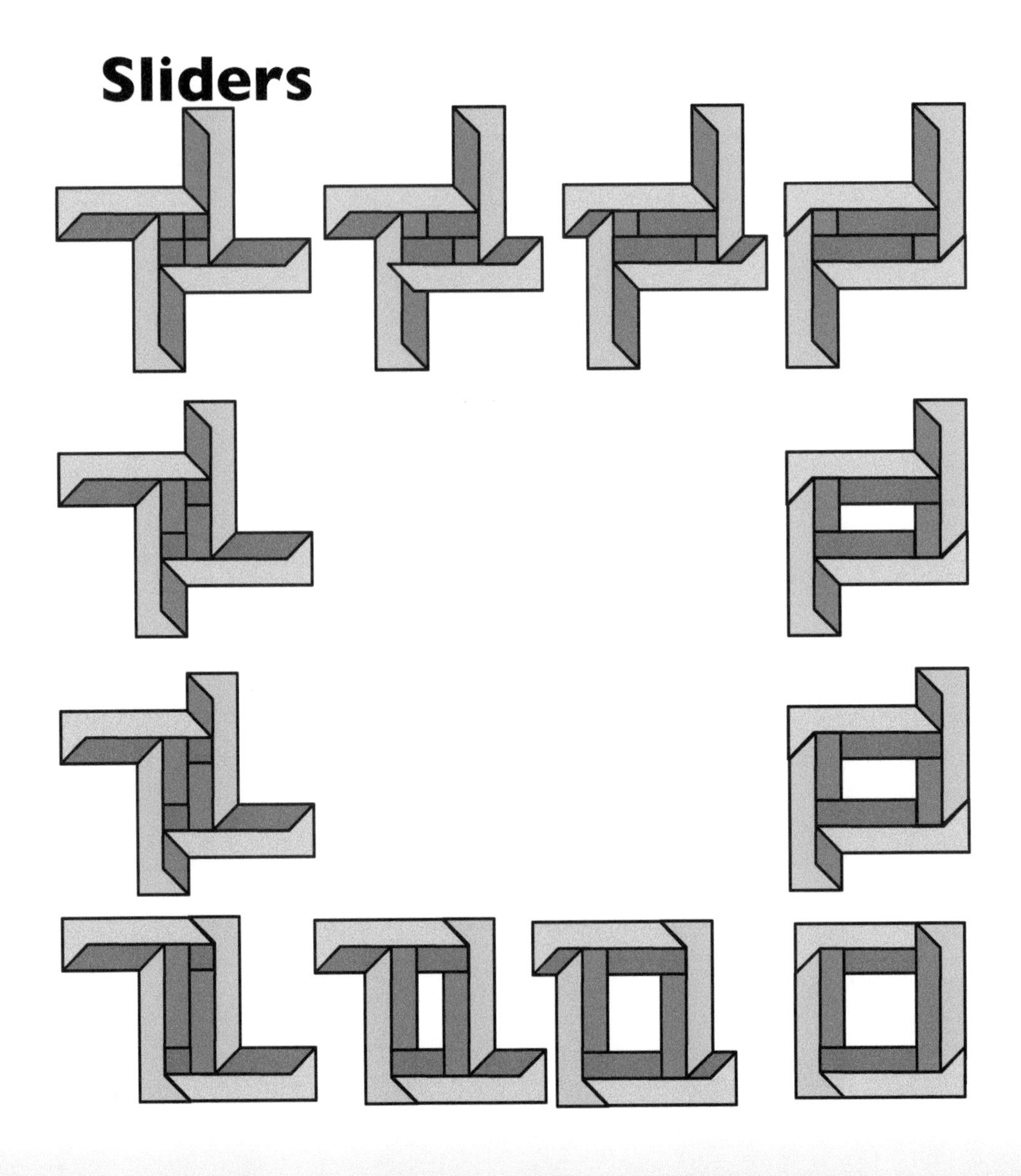

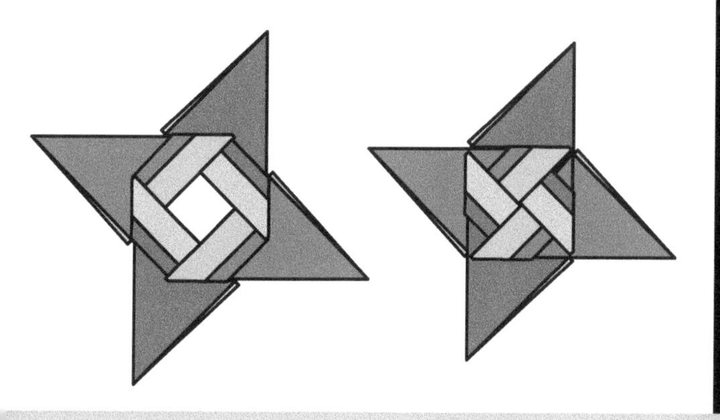

Star 4 Slider
Square From Silver Rectangle

★

Use four silver rectangles
or ratio between 1.25:1 and 1.5:1
Finished model ratio 2.5

The first model in this book, *Star 4, Square From Silver Rectangle*, can slide a little. This version enhances the sliding action by making it longer and more rigid.

Module ★

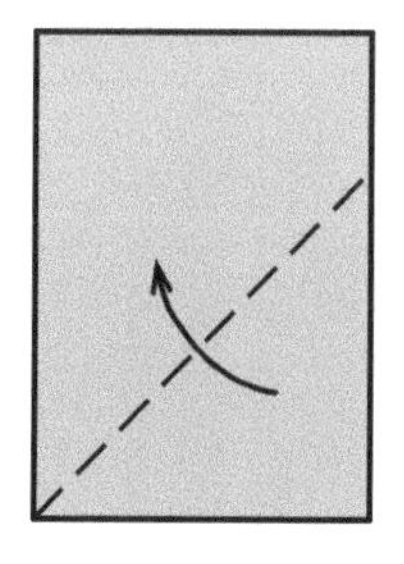

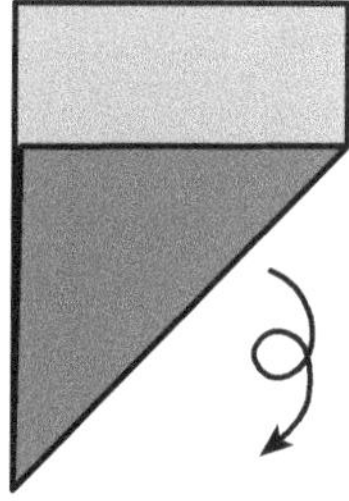

1 Take the lower right corner and fold the diagonal of a square. Turn over.

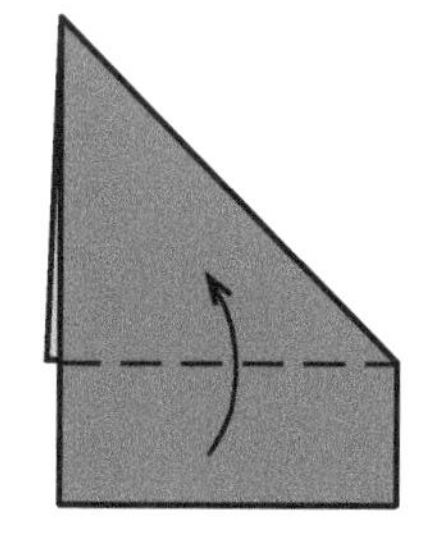

2 Fold the lower flap up so that the fold aligns with the raw edge behind it.

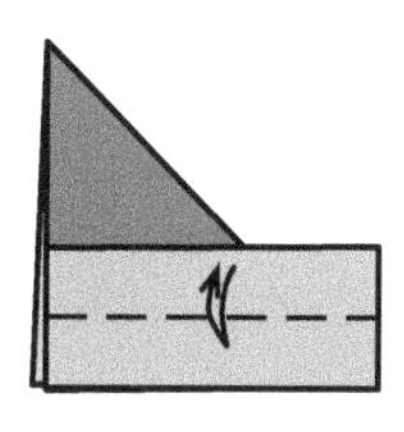

3 Fold the top edge of the leftover rectangle to the bottom edge and unfold.

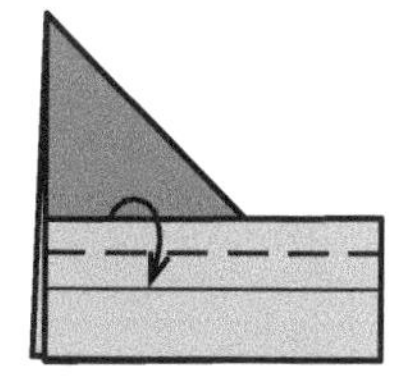

4 Fold one quarter of the leftover rectangle down.

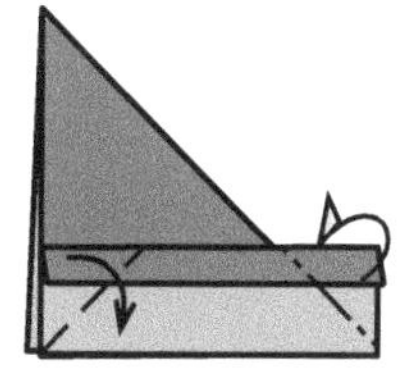

5 Make two 45° folds on the new flap. Leave a small gap for the mountain fold so that it will slide.

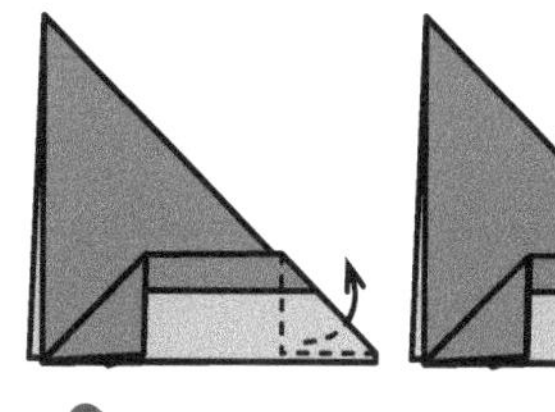

6 Unfold the right triangular flap.

Assembly ★

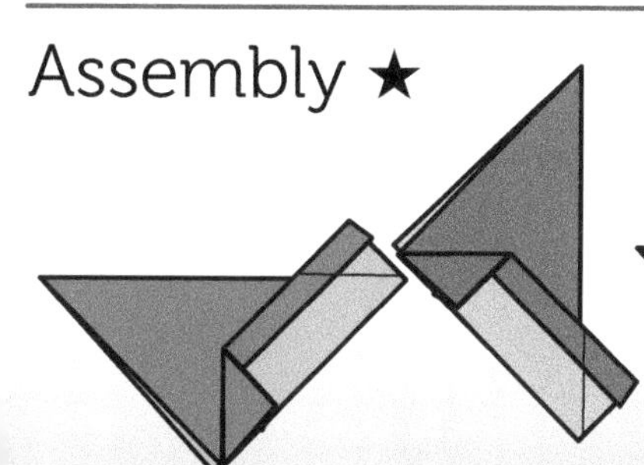

1 Hook the second unit onto the first.

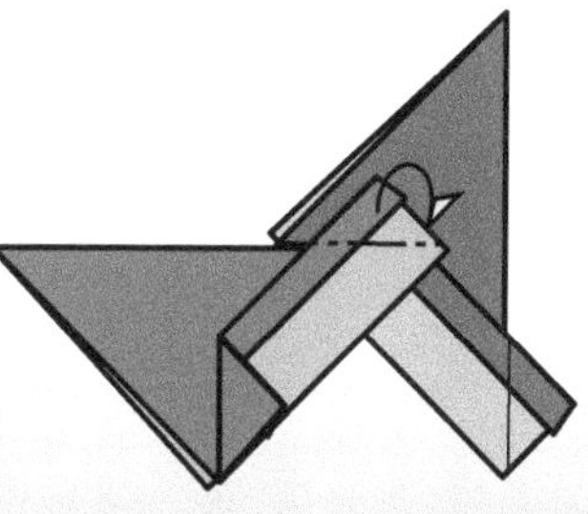

2 Mountain fold the flap of the first unit.

3 Add more units.

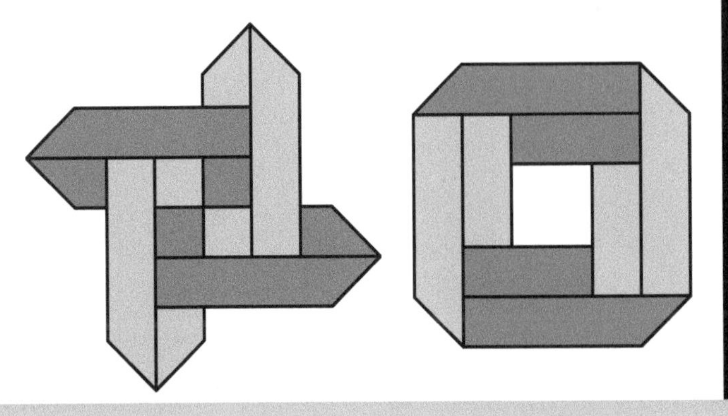

Square Pointer Slider

★★

Use four silver rectangles A7 or ratio more than 1.2:1
Finished model ratio 1.6

This model can be made from oblongs other than silver rectangles. When folded from a square, the units will not slide. This unit is known as a *pig base* because it is the basis for the traditional pig. It also makes Dave Mitchell's action model *Ad Infinitum* [Lam and Pope 16].

Module ★★

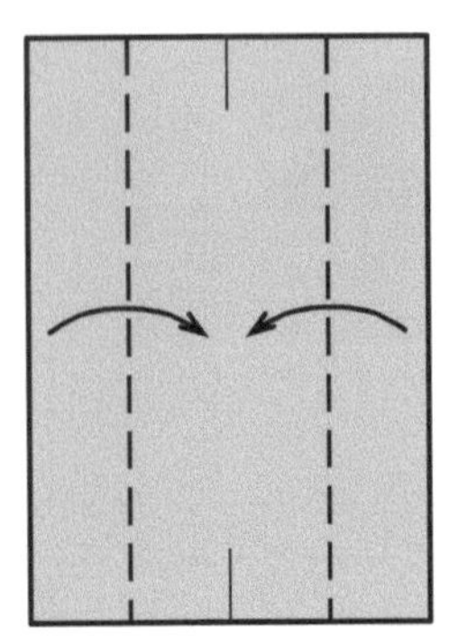

1 Pinch the midpoints of the short edges. Fold the long sides to the middle.

2 Crease the corners.

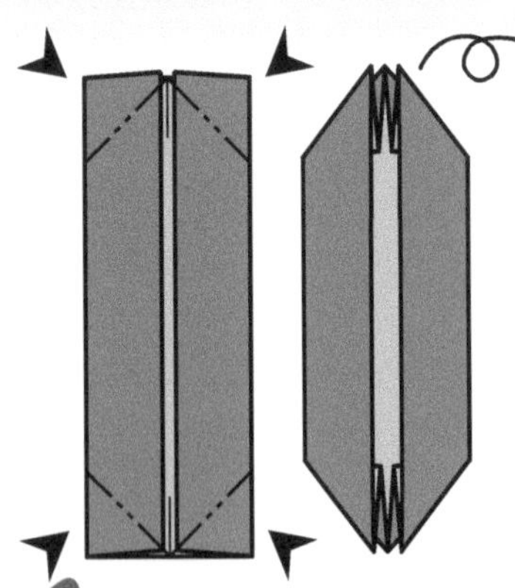

3 Inside reverse fold the corners. Turn over.

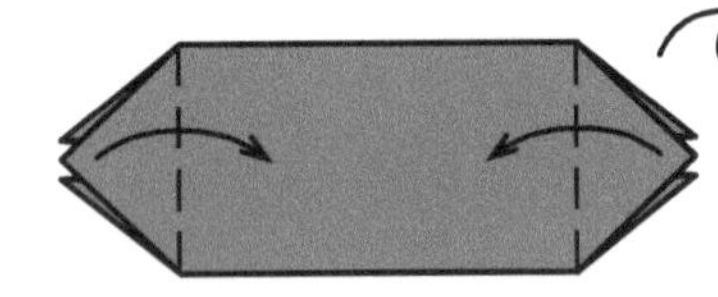

4 Fold the flap inwards.

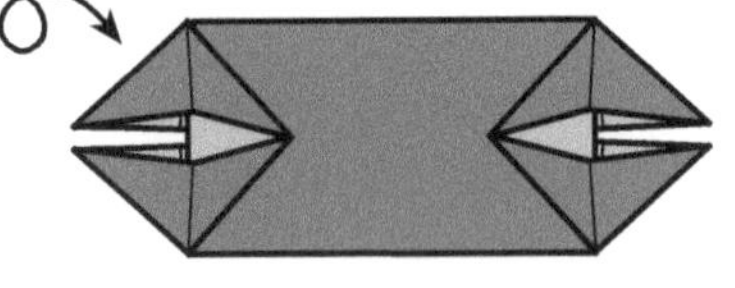

5 Turn over.

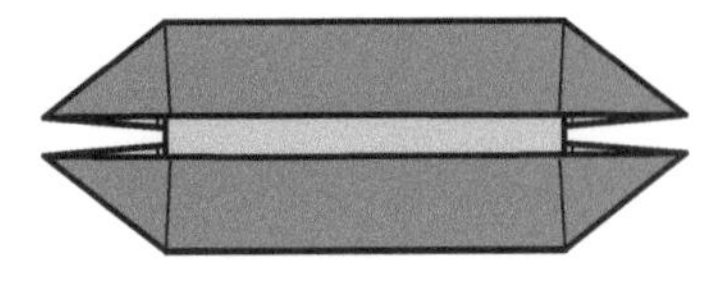

6 Unit complete.

Assembly ★★

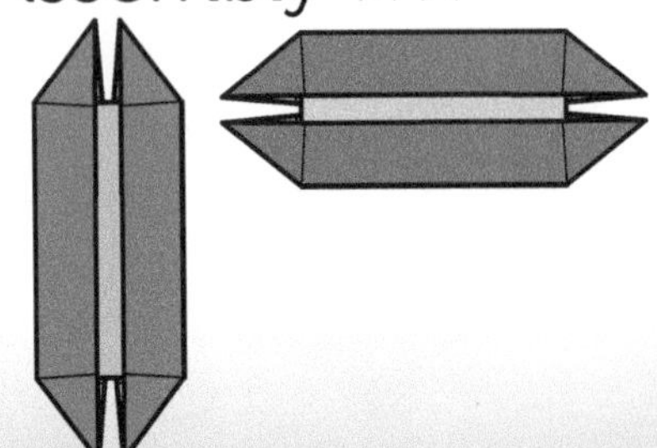

1 Insert the first unit into the second.

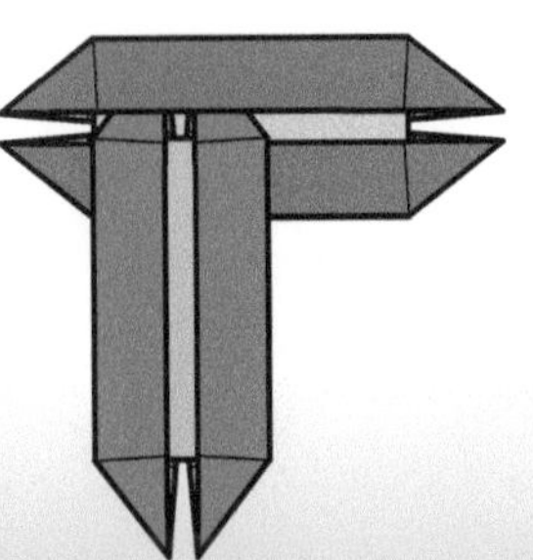

2 Add more units in a clockwise direction.

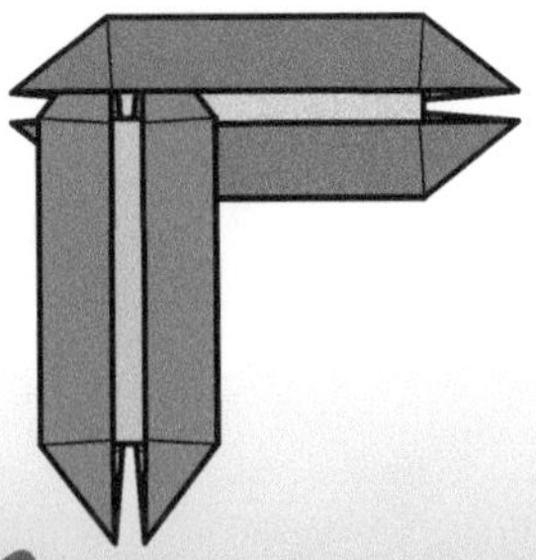

3 Slide the units to make space if needed.

Pinwheel Square Slider

★★
Use four 75 mm squares
Finished model ratio 1.4

This is a simple and effective slider that can work as a pinwheel. Put a pencil through the hole and blow from the side.

Module ★★

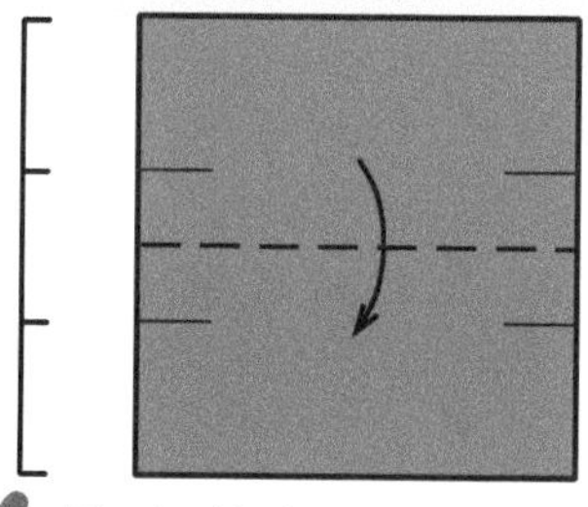

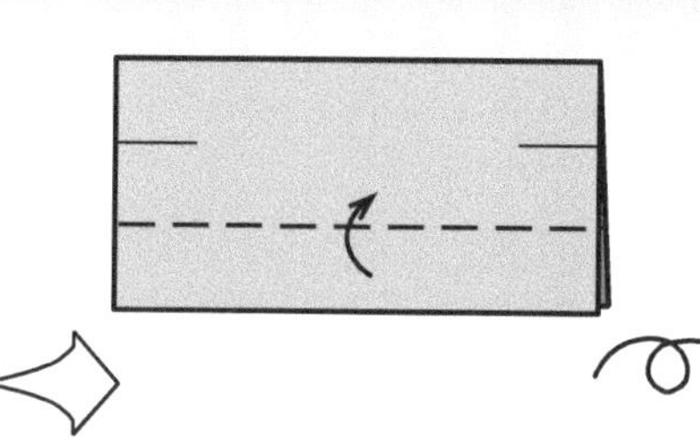

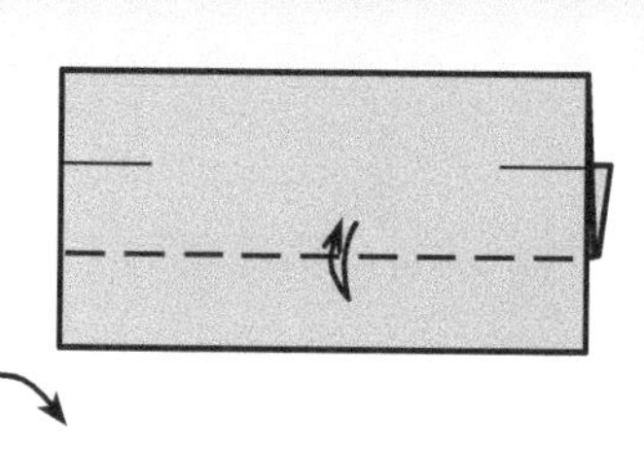

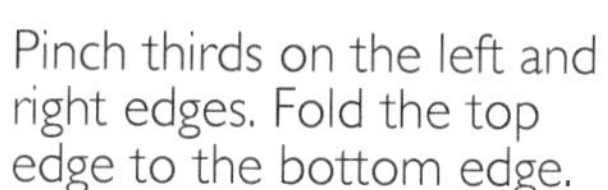

1 Pinch thirds on the left and right edges. Fold the top edge to the bottom edge.

2 Fold up the bottom edge to the pinches. Turn over.

3 Fold up the bottom edge to the pinches and unfold.

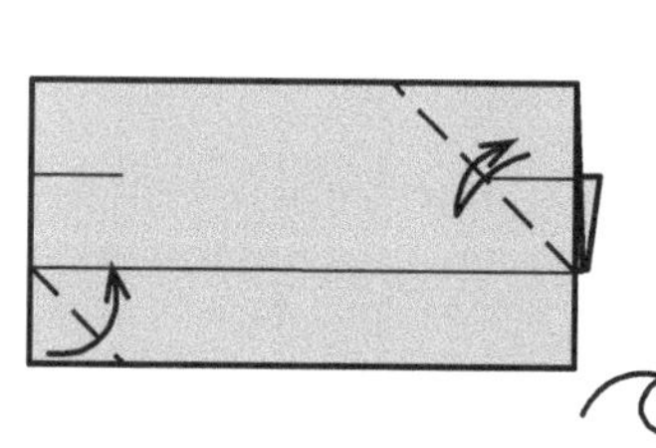

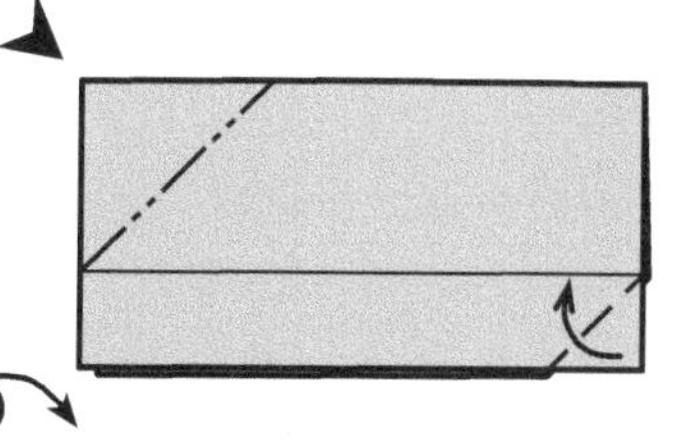

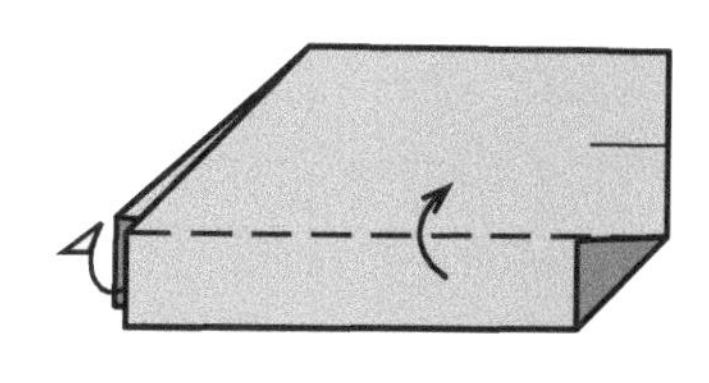

4 Fold the bottom left corner. Crease the top right corner. Turn over.

5 Fold the bottom right corner to match behind. Inside reverse fold the top left corner.

6 Fold up the bottom edges.

Assembly ★★

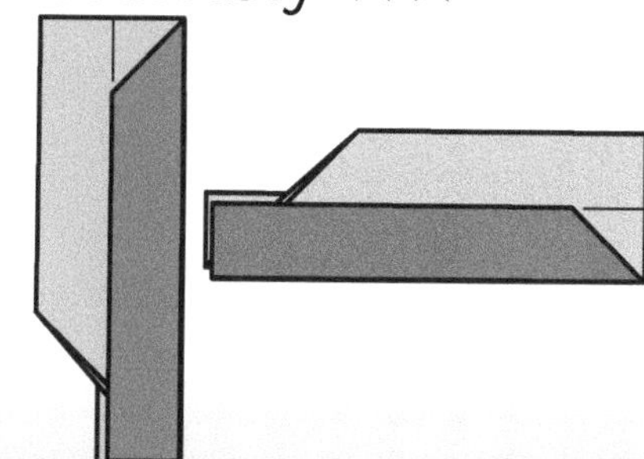

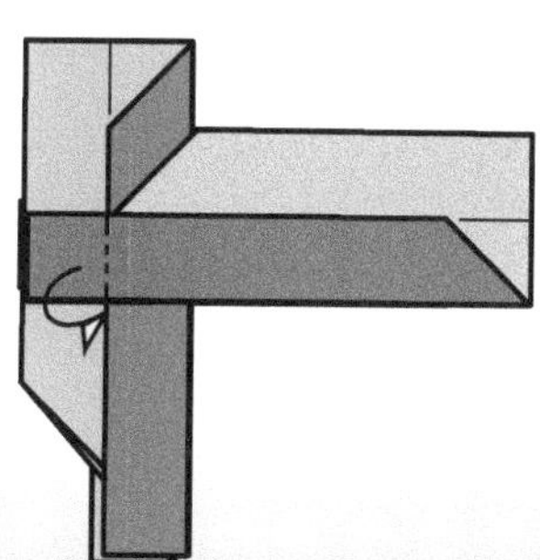

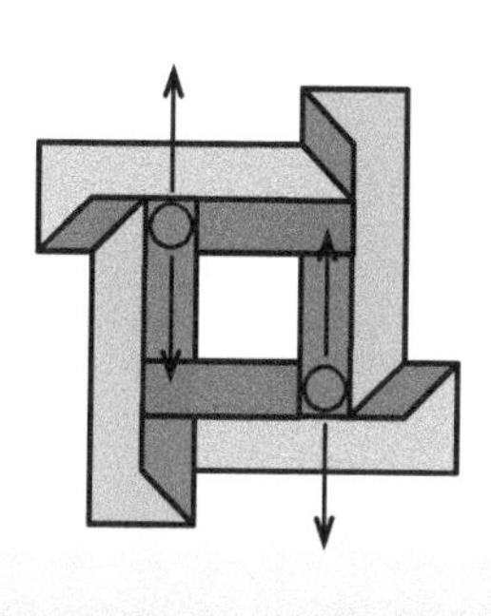

1 Insert the first unit between the arms of the second unit.

2 Mountain fold the tips of the arms of the second unit into the first unit.

3 Add units in a clockwise direction. Slide to make space, if needed.

Pinwheel Square Slider

★★★
Use four 75 mm squares
Finished model ratio 1.4

Inserting a strip of paper in the previous model makes a 3D version: this technique is known as grafting. You can apply this technique to other models that have a plane of symmetry.

Module ★★★

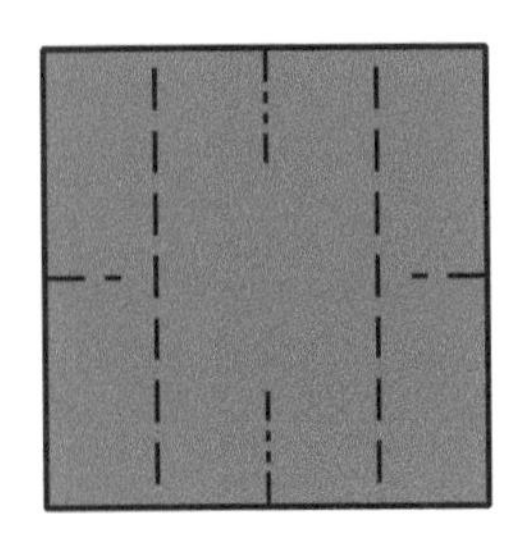

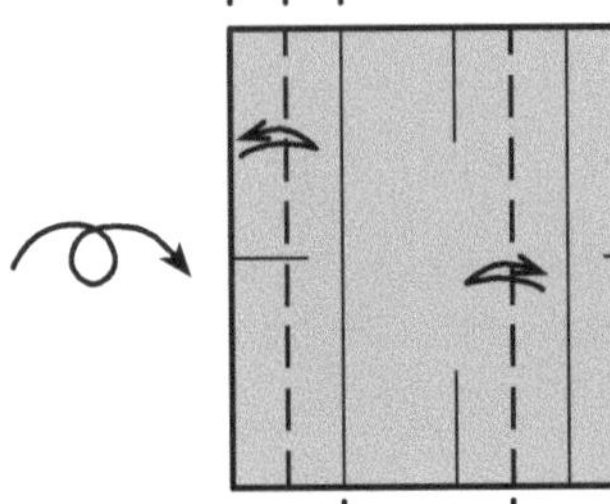

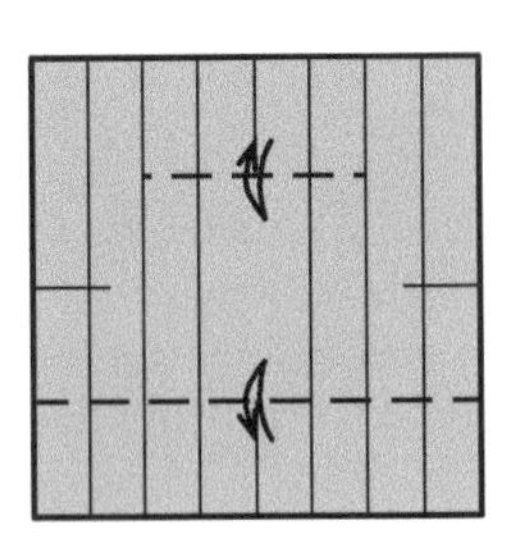

1 Pinch midpoints of each edge. Fold the left and right edges to the centre and unfold.

2 Fold the left and right edges to the quarter line. Unfold. Repeat with the three quarter line.

3 Fold the bottom edge to the middle. Fold the top edge to the middle but only crease the central portion. Unfold.

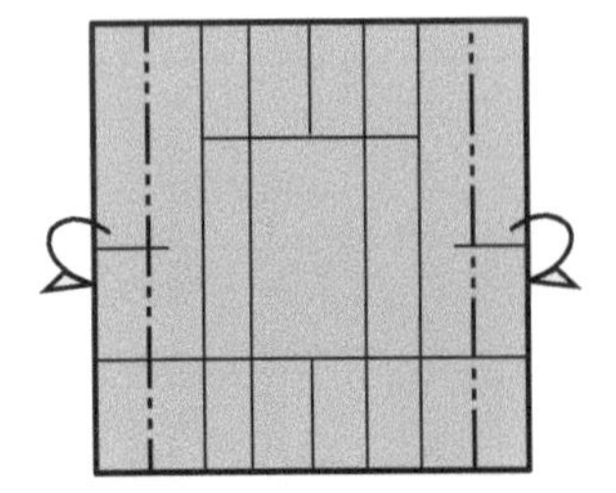

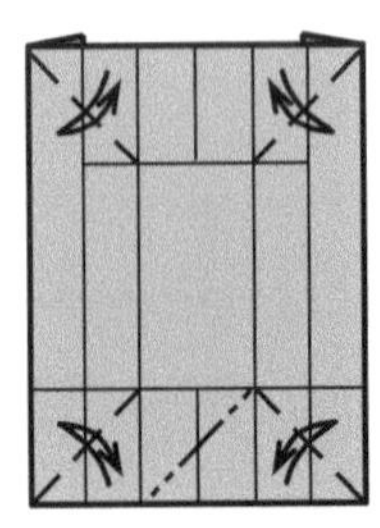

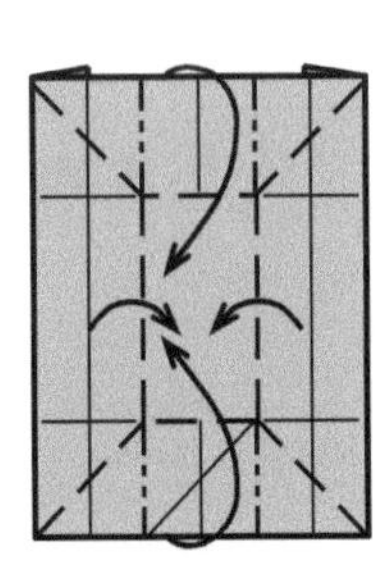

4 Fold one eighth behind on the left and right sides.

5 Crease the upper layer only.

6 Form a box shape.

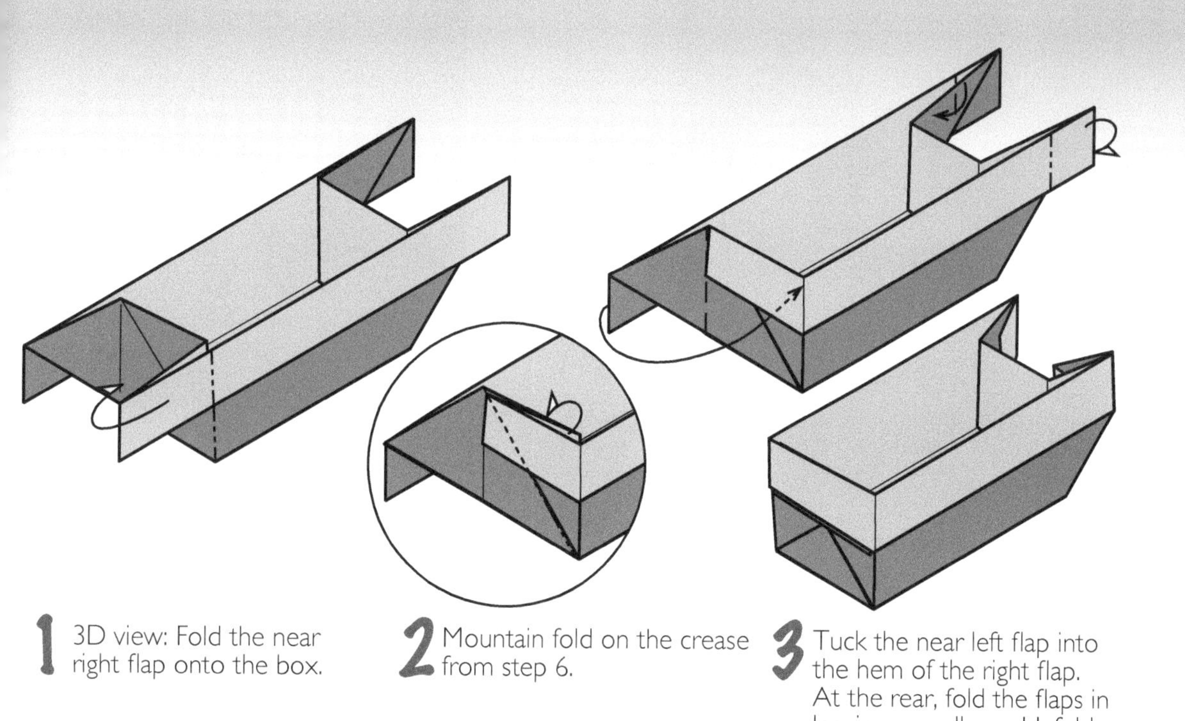

1 3D view: Fold the near right flap onto the box.

2 Mountain fold on the crease from step 6.

3 Tuck the near left flap into the hem of the right flap. At the rear, fold the flaps in leaving a small gap. Unfold.

Assembly ★★

1 Insert the second unit into the arms of the first.

2 Mountain fold the tips of the arms of the first unit into the second.

3 Join the other two units.

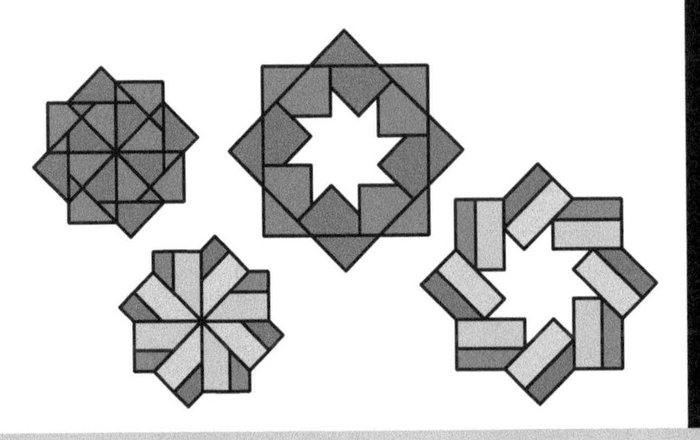

Octagram Slider

★

Use eight 1:2 rectangles
Finished model ratio 2.0

This is the Octagram Wreath with a different location point. The sequence finds this kite point with a fold that is hidden in the final unit. Both sides of the slider are attractive.

Module ★

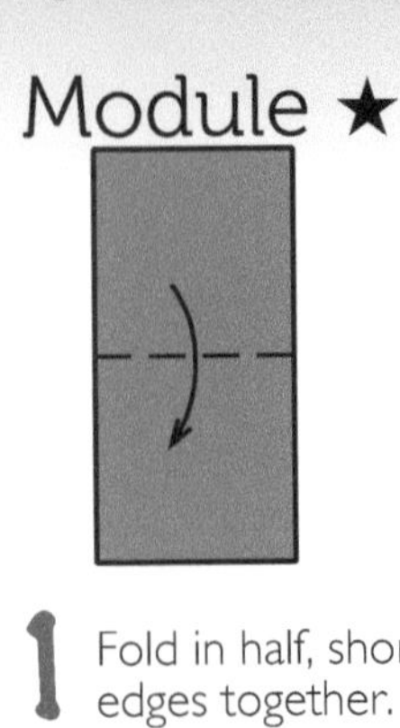

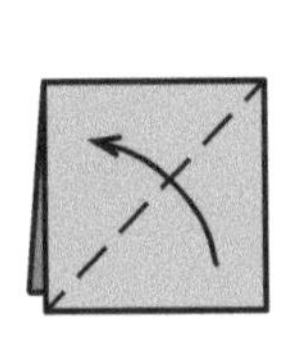

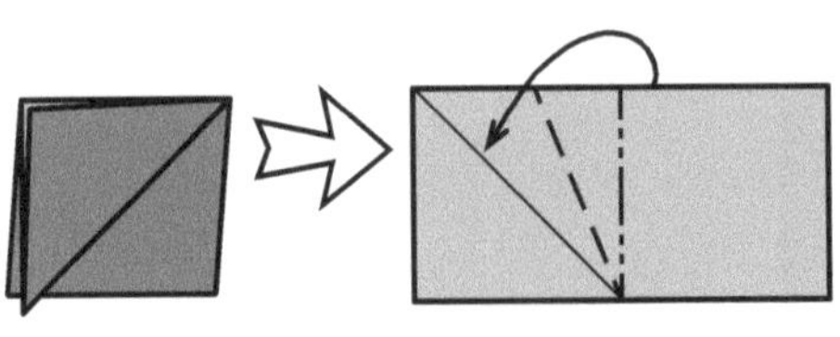

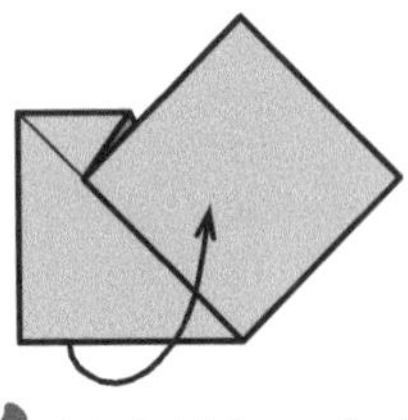

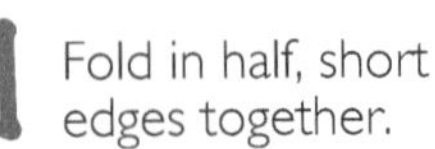

1 Fold in half, short edges together.

2 Fold the diagonal. Unfold.

3 Radial pleat the central line onto the diagonal.

4 Unfold from behind. Let the flap flip forwards.

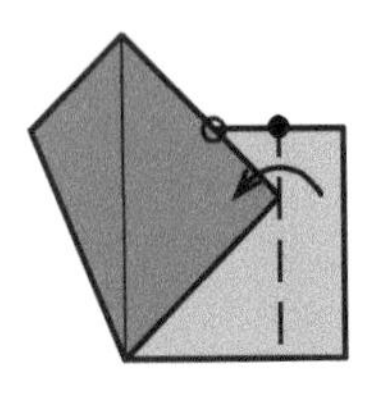

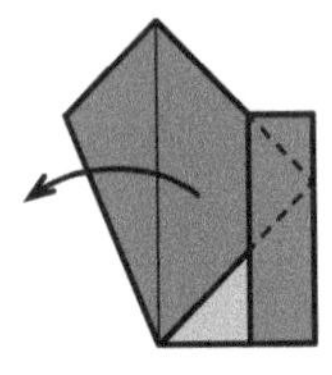

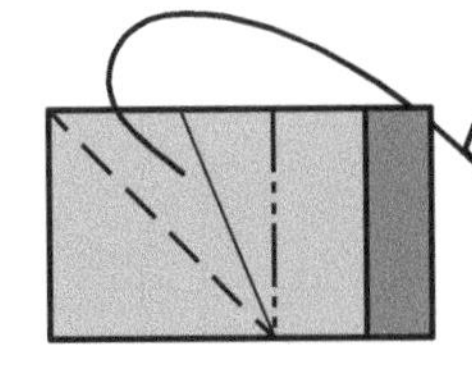

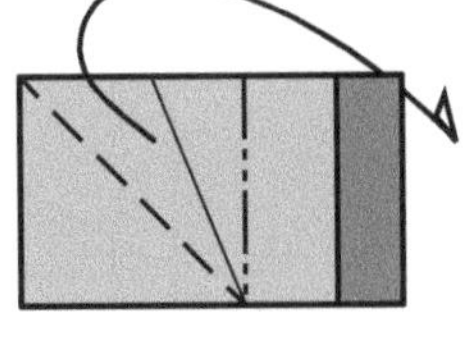

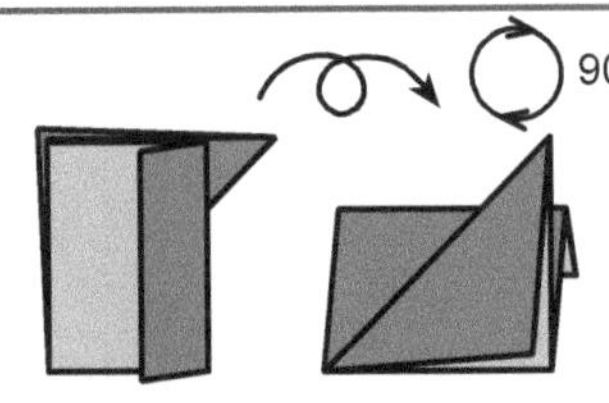

5 Fold the right edge over the corner.

6 Unfold the left flap.

7 Radial pleat.

8 Turn over and rotate.

Assembly ★

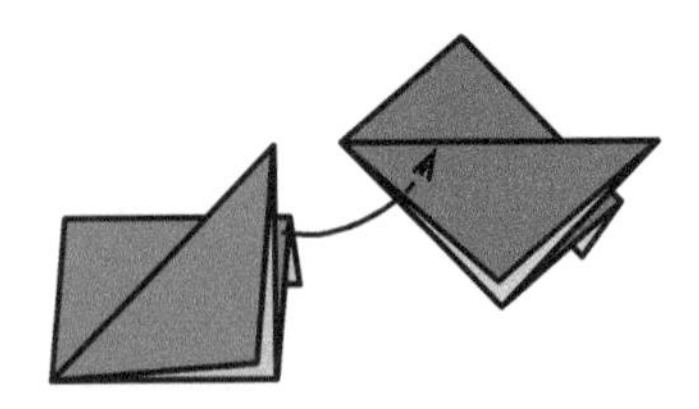

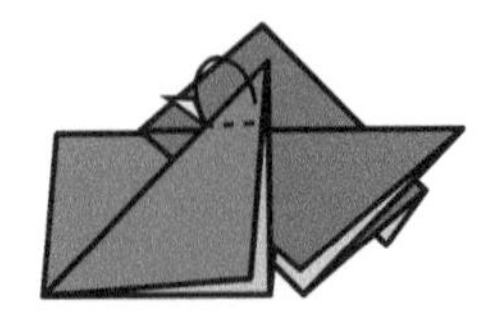

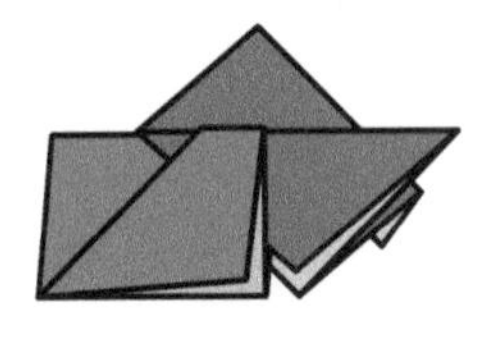

1 Insert the rear flap of the first unit between the upper two layers of the second unit.

2 Mountain fold the tip of the first unit to lock.

3 Add more units in a clockwise direction.

Floppy Slider 8

★

Use eight 75 mm squares
Finished model ratio 2.0

The units of the sliders so far have clearly defined paths. Although there is a reasonable amount of play in this slider, the units form specific configurations. Smooth paper like origami paper may help the units slide.

Module ★

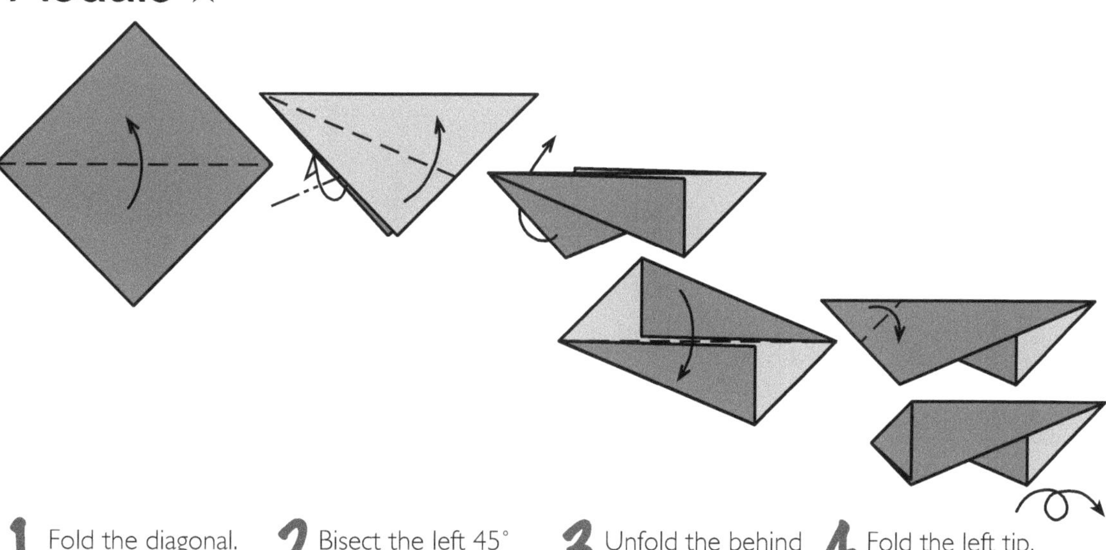

1 Fold the diagonal.

2 Bisect the left 45° angle. Turn over and repeat.

3 Unfold the behind and valley fold.

4 Fold the left tip. Turn over.

Assembly ★

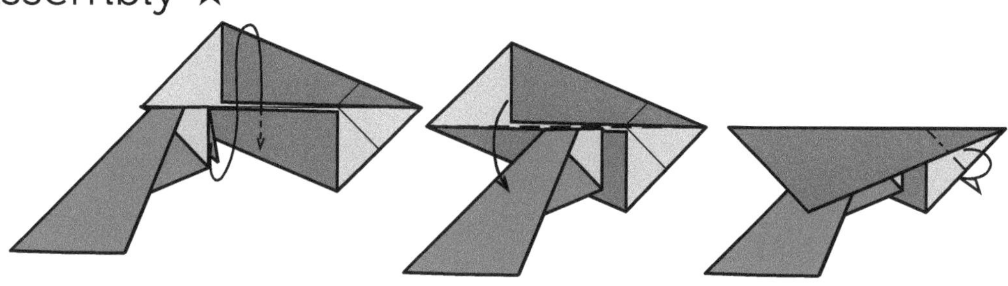

1 Unfold the second unit. Hook the tip of the first unit.

2 Fold down the top flap of the second unit to lock.

3 Fold the tip of the second unit so that it will hook into the third unit.

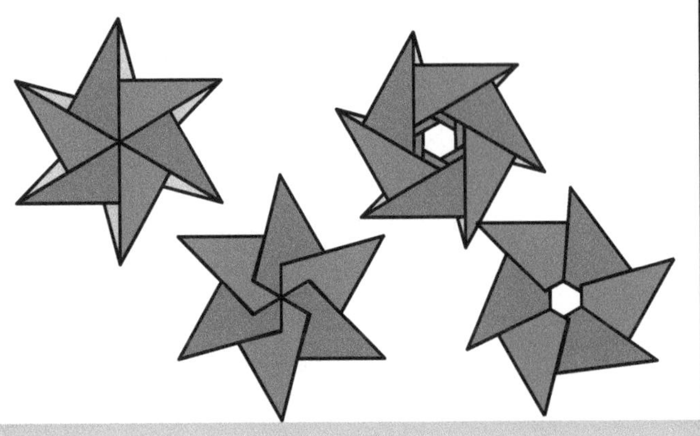

Floppy Slider 6

★

Use eight 75 mm squares
Finished model ratio 1.6

This is a six-piece version of the previous model. The flaps made in module step 2 can catch when sliding, so flip them out of the way if needed.

Module ★

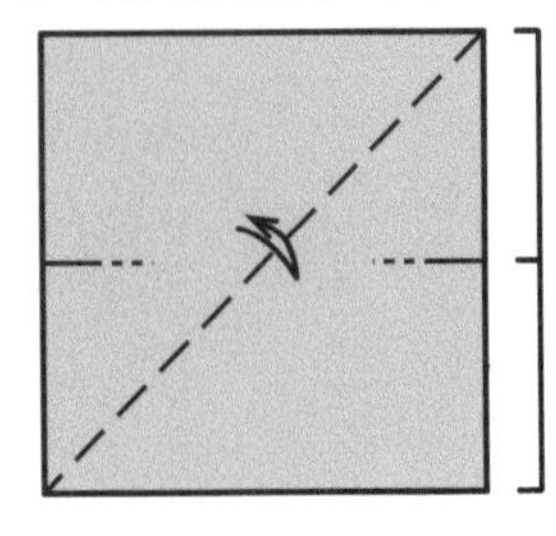

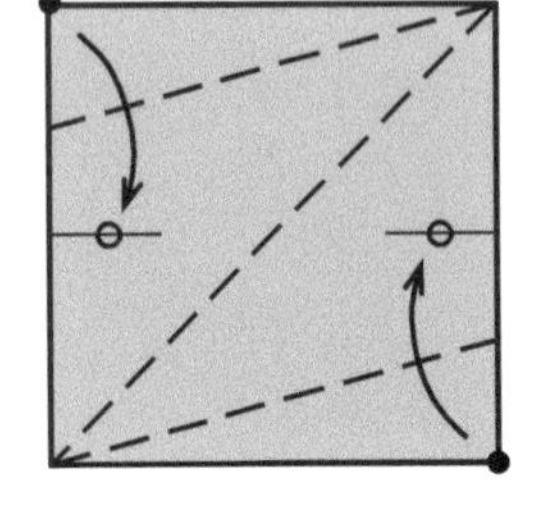

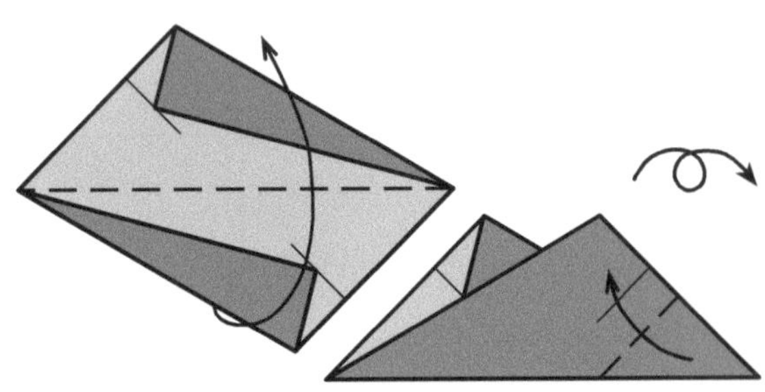

1 Crease the diagonal. Crease in half without folding the central portion.

2 Make a 15° fold. Rotate 180° and repeat.

3 Fold the diagonal.

4 Fold the right tip. Turn over.

Assembly ★

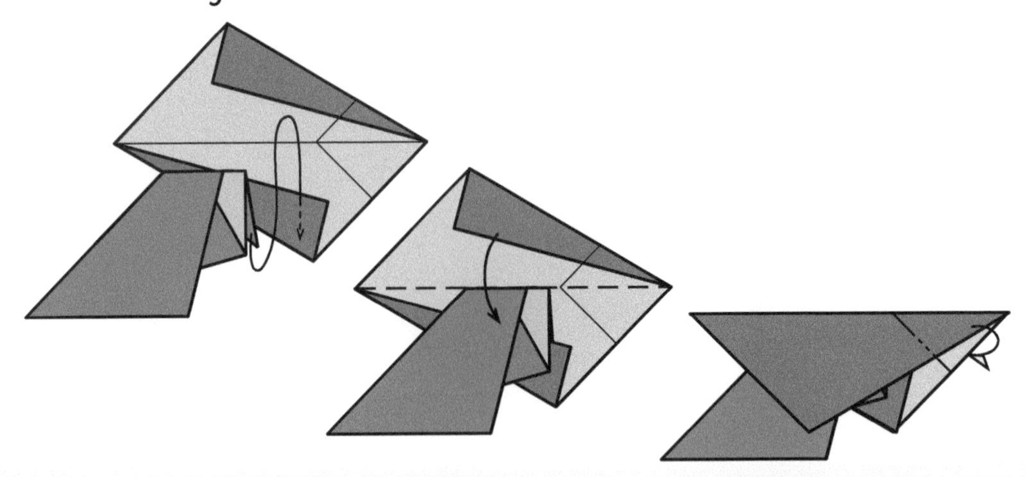

1 Unfold the second unit. Hook the tip of the first unit.

2 Fold down the top flap of the second unit to lock.

3 Fold the tip of the second unit so that it will hook into the third unit.

Slider N 8

★★
Use eight 75 mm squares or larger
Finished model ratio 1.5

Mono or duo paper works well for this model. Assembly steps 3 and 4 show two different ways that the units can be joined: both slide well. You may want to use one of the units as a template for making thirds. You can use the template as the last module.

Module ★★

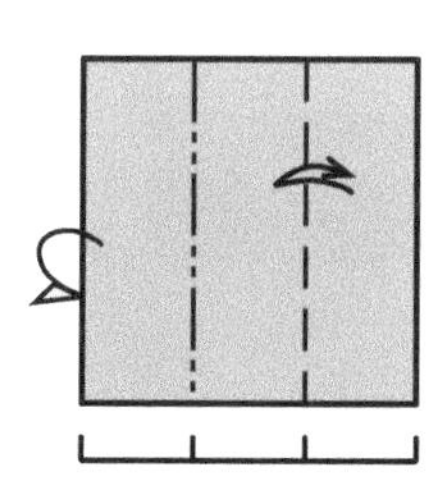

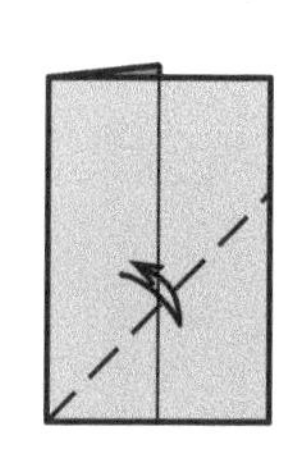

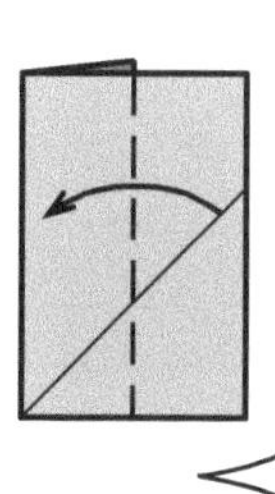

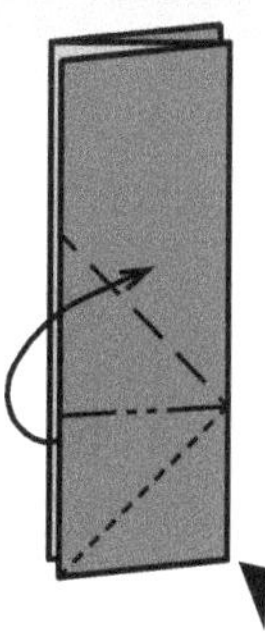

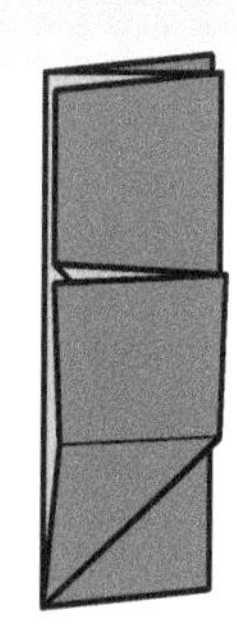

1 Pleat into thirds and unfold. Mountain fold the left third.

2 Bisect the lower left right angle and unfold. Fold the right flap over.

3 Swivel the paper upwards and squash the flap.

4 Unit complete.

Assembly ★★

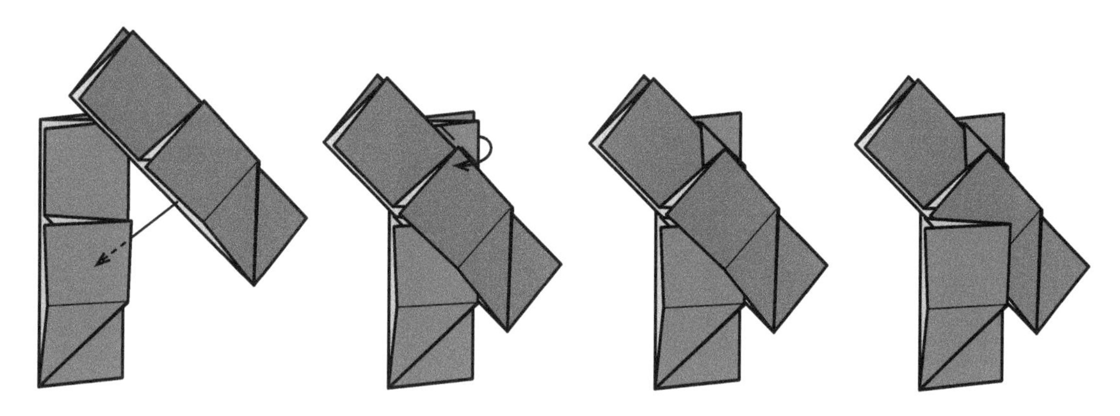

1 Interleave the two units.

2 Only fold the front flap. Make a slightly loose fold so that the units will slide.

3 Swivel the paper upwards and squash the flap.

4 The flaps can be pulled out to show a different pattern.

Slider N 6

★★
Use six 75 mm squares or larger
Finished model ratio 1.4

This is a six-piece version of the previous model. Each unit uses another as a template for finding the width needed for sliding. As with the other *Slider N* models, try using paper that is the same colour on both sides, or has different colours.

Module ★★

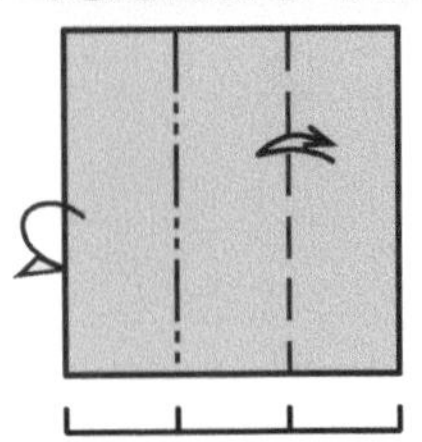

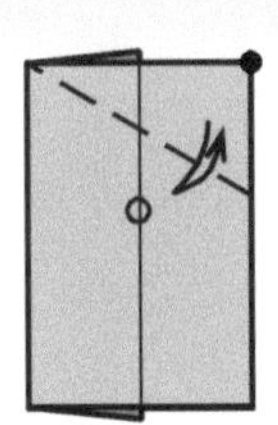

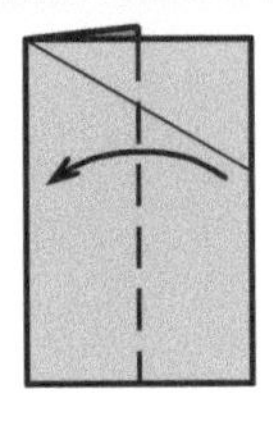

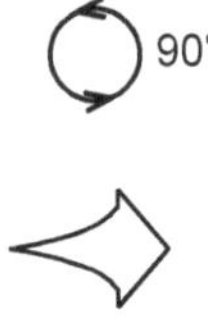

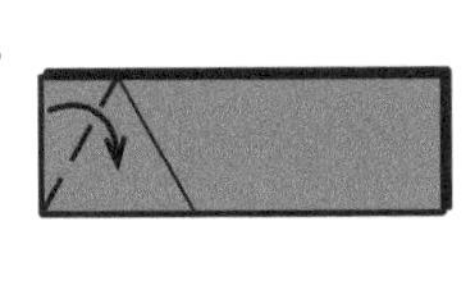

1 Pleat into thirds and unfold. Mountain fold the left third.

2 Crease a 60° at the top and unfold. Fold the right flap over.

3 Fold on the existing crease.

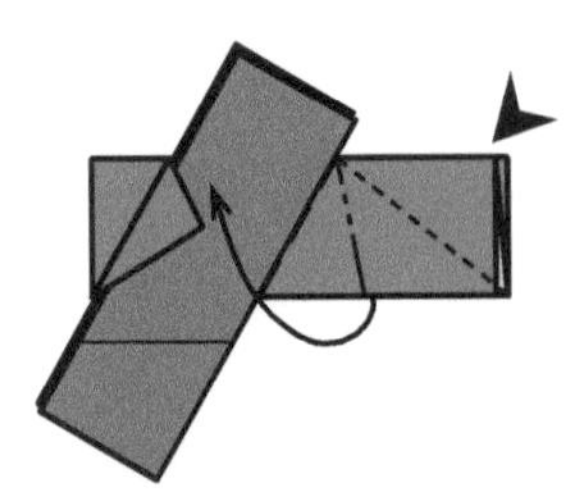

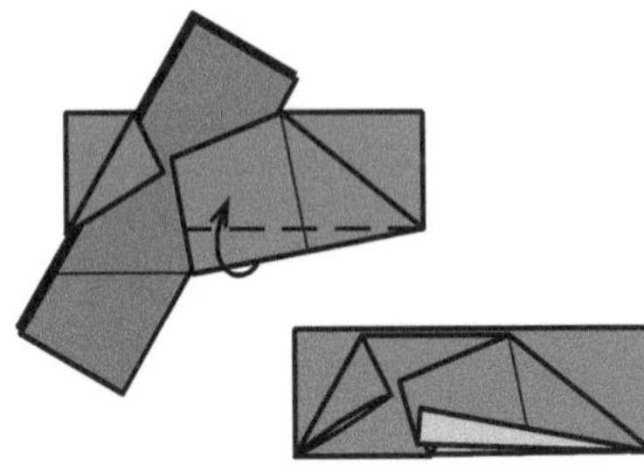

4 Insert a second unit under the new flap.

5 Take the top layer and fold over the template. Swivel the paper across and squash the flap.

6 Fold the edge up. Remove the template.

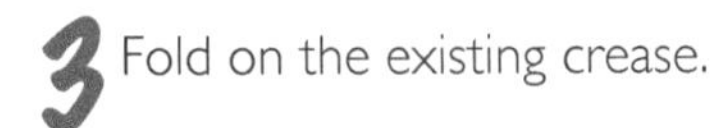

Assembly ★★

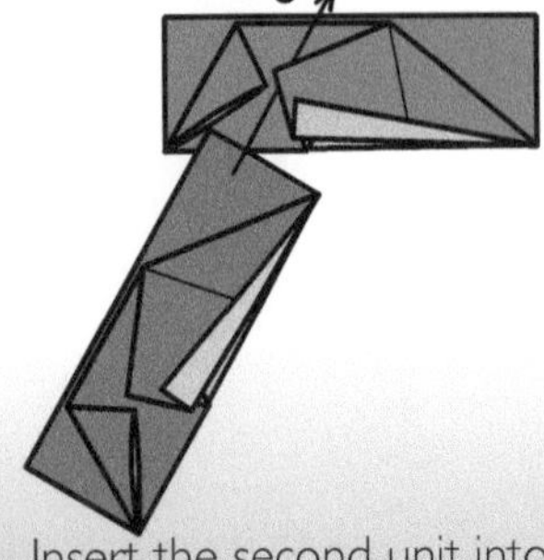

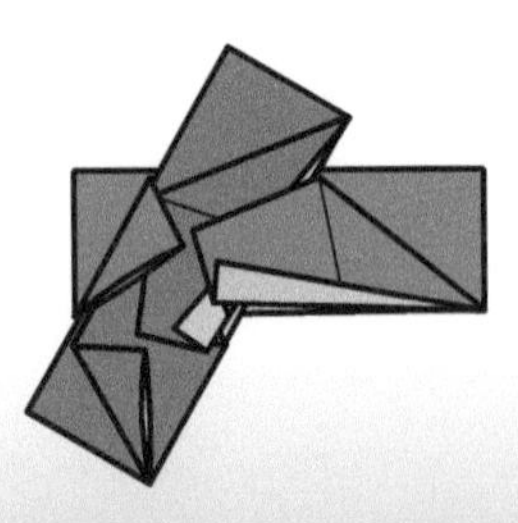

1 Insert the second unit into the first.

2 Tuck the flaps of the second unit into the first.

3 Add more units. Slide them to make space if needed.

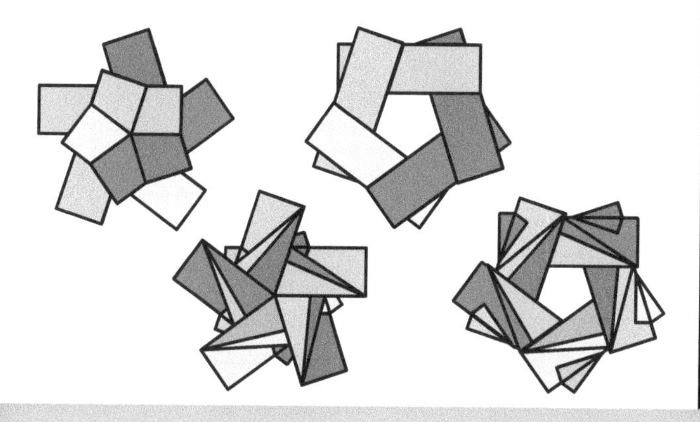

Slider N 5

★★
Use five 75 mm squares or larger
Finished model ratio 1.4

It may seem surprising that an odd number of units can slide. However, holding two points and moving them makes anther unit move, one that is not directly held. Notice how the unit achieves the 1:3 approximation of 18°.

Module ★★

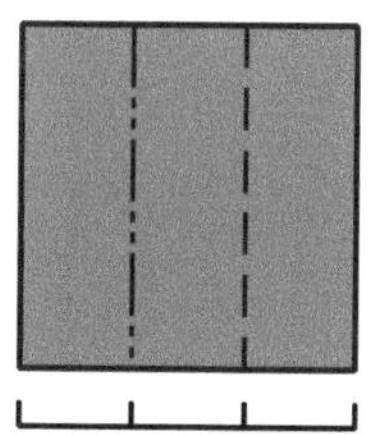

1 Pleat into thirds and unfold.

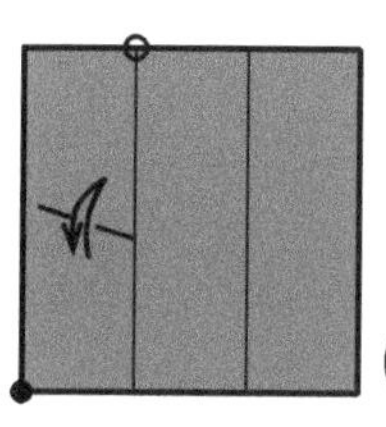

2 Crease the bottom left corner to one third along the top. Reform pleats.

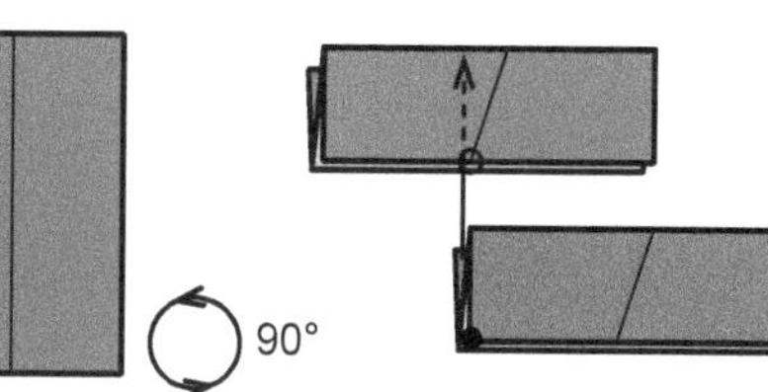

3 Make a second unit and put it in the first.

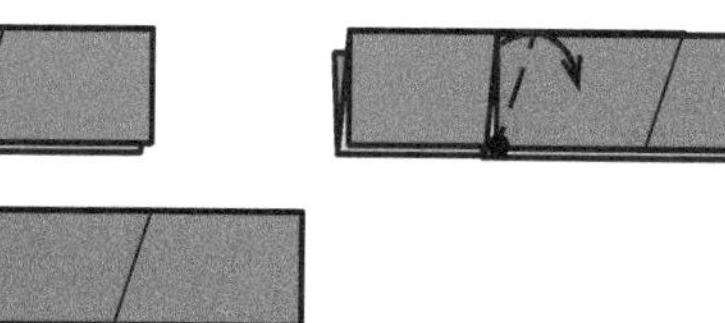

4 Fold the second unit along the crease of the first unit.

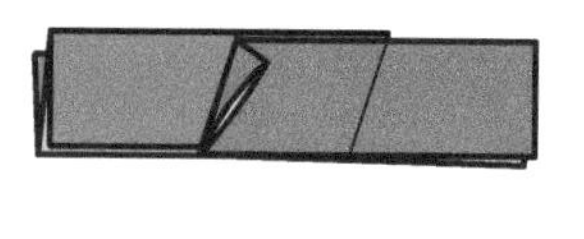

5 Remove the first unit. Put it under the folded flap.

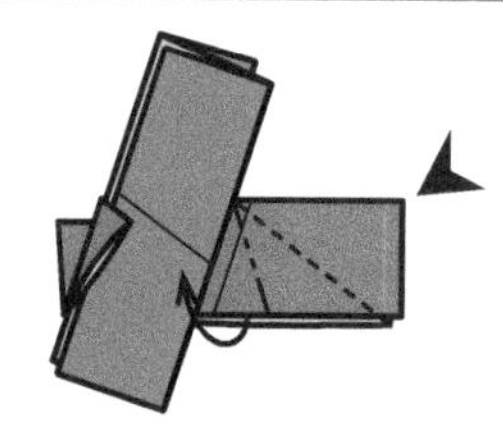

6 Fold the top layer of the second unit against the first unit and squash.

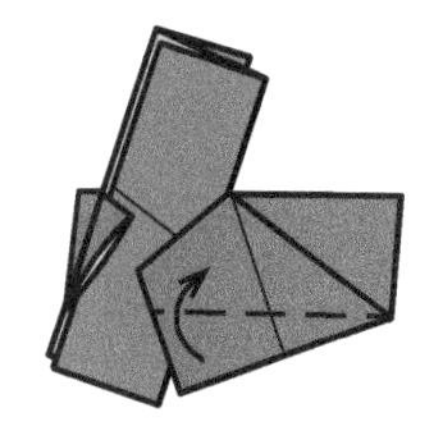

7 Fold up.

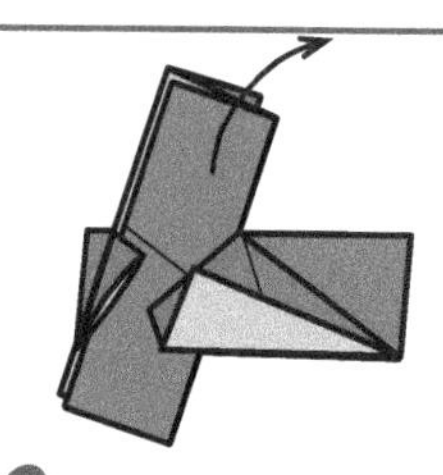

8 Remove the first unit.

Assembly ★★

1 Insert the first unit into the second.

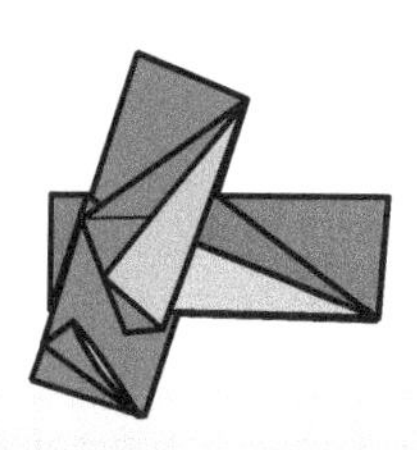

2 Tuck the flaps of the first unit into the second.

Chapter 11

3D Stars

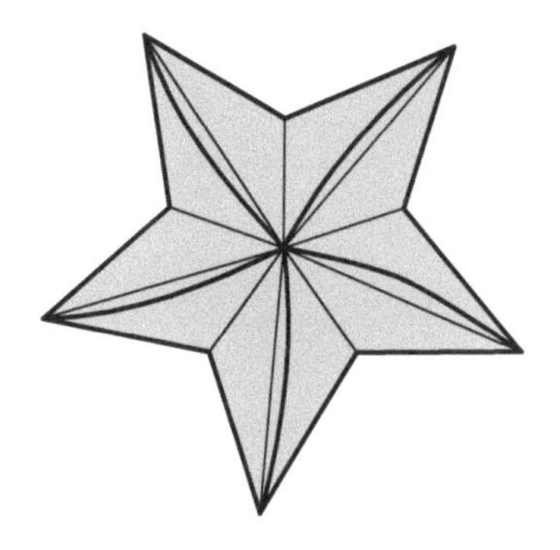

Starfish

★

Use five A8 rectangles or larger. You can also use longer rectangles and a different number of units. Finished model ratio 1.2

This folding method generalises to rectangles that are longer than silver rectangles. For bronze rectangles there is no excess paper to fold in steps 4 and 7 as the right angle is trisected. Longer rectangles (e.g. 2:1) have the excess paper on the lower right instead of the centre in step 4.

Module ★

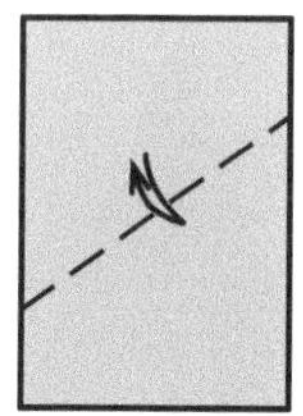

1 Fold the lower right corner to top left corner and unfold.

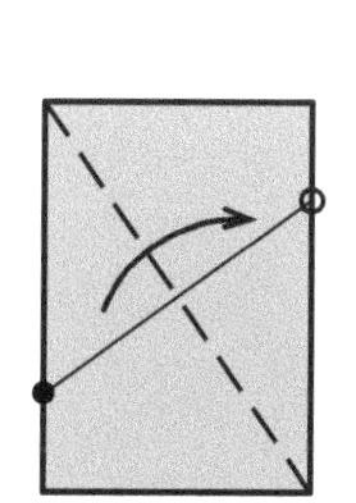

2 Fold the diagonal by folding the left end of the crease line onto the right end.

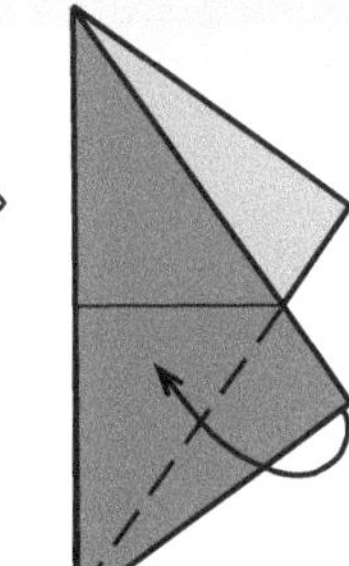

3 Fold the lower right flap to align with the raw edge behind.

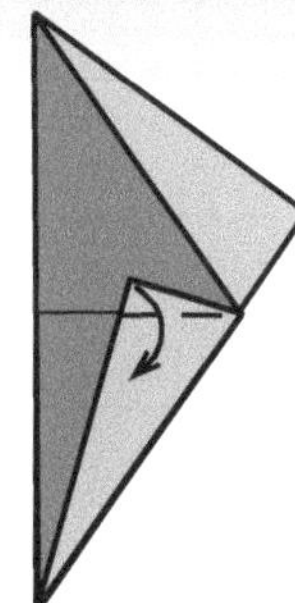

4 Fold the tip of the flap down along the crease made in step 1.

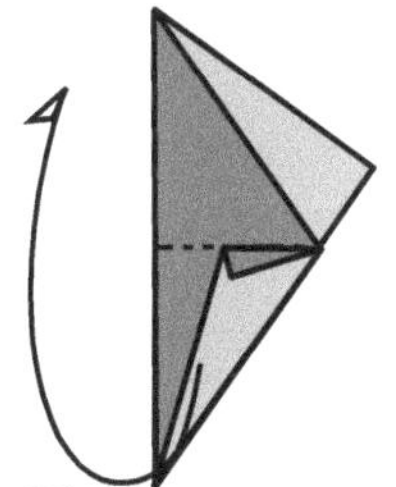

5 Fold in half by mountain folding the bottom behind.

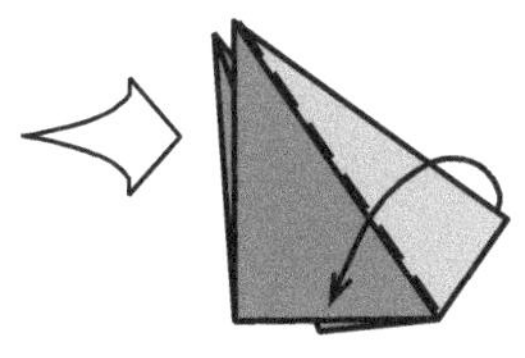

6 The next steps are similar to steps 3 and 4. Fold the right flap over the raw edge.

7 Fold the tip of the right flap up to match the folded edge behind.

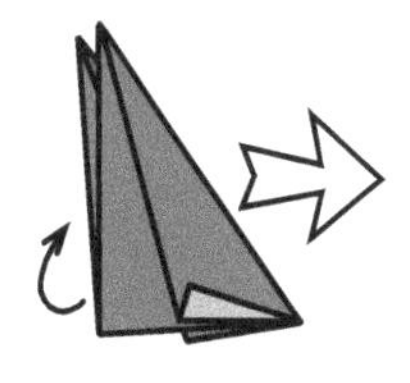

8 Open the paper and return to step 3.

Assembly ★

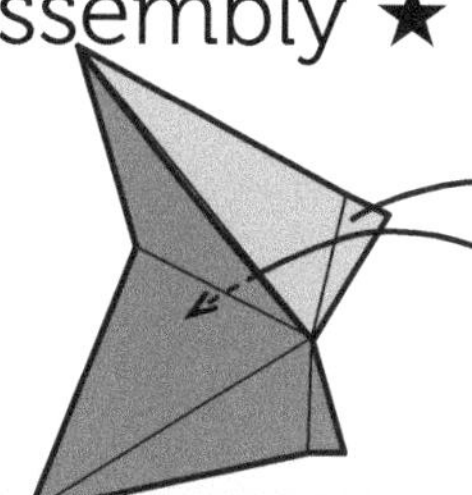

1 Slide the flap of each unit into the pocket of the other unit.

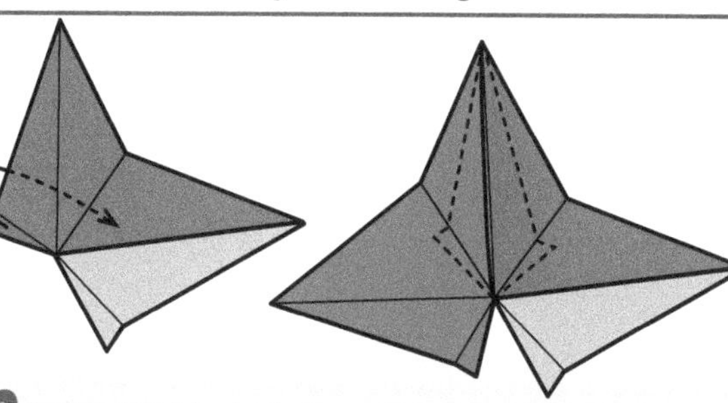

2 Valley fold to lock the units together. Add more units.

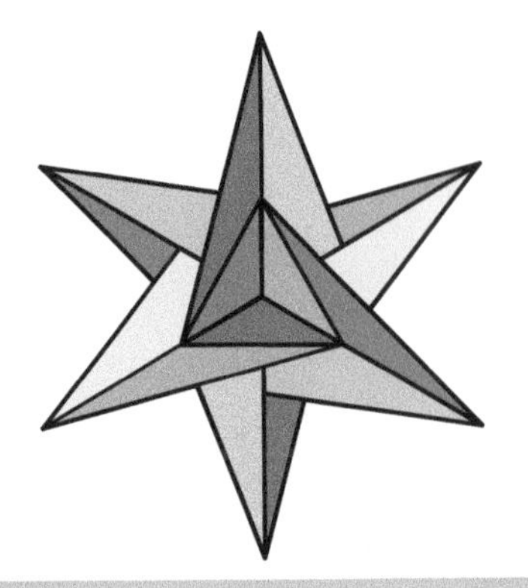

XYZ Rhombic

★

Use six 2:1 rectangles,
e.g. cut 100 mm squares
Finished model ratio 2.0

This simple module works well from paper that is the same colour on both sides. The name comes from the three mutually perpendicular planes of analytic geometric, i.e. 3D Cartesian coordinates like (1, 0, 0).

Module ★

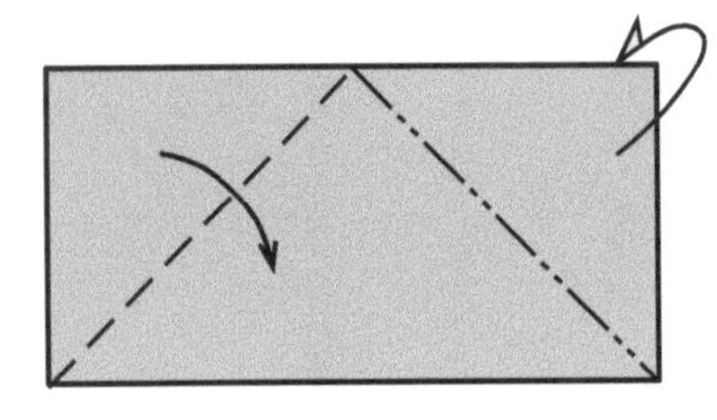

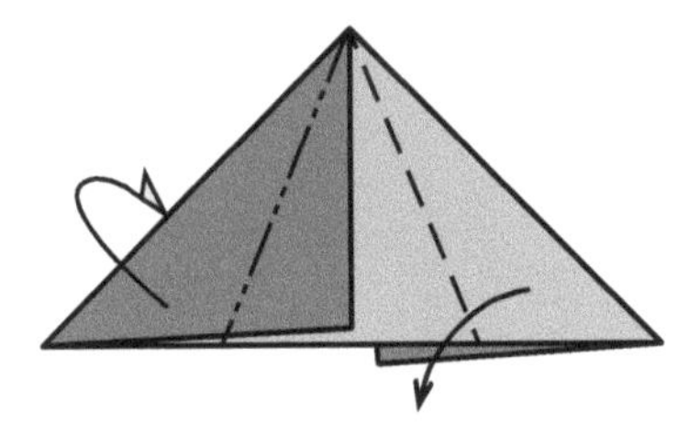

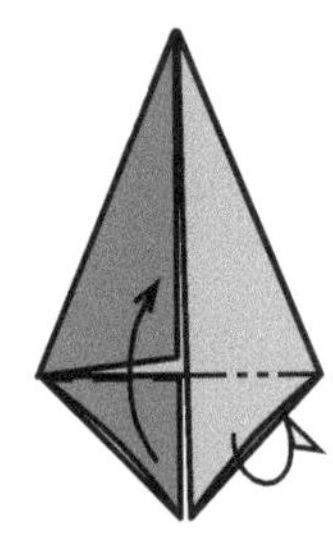

1 Bisect the lower left right angle. Turn over and repeat.

2 Bisect the 45° angles at the top. Fold so that the original corner of the rectangle is on the outside.

3 Fold the bottom flaps over the original corners of the square.

Assembly ★

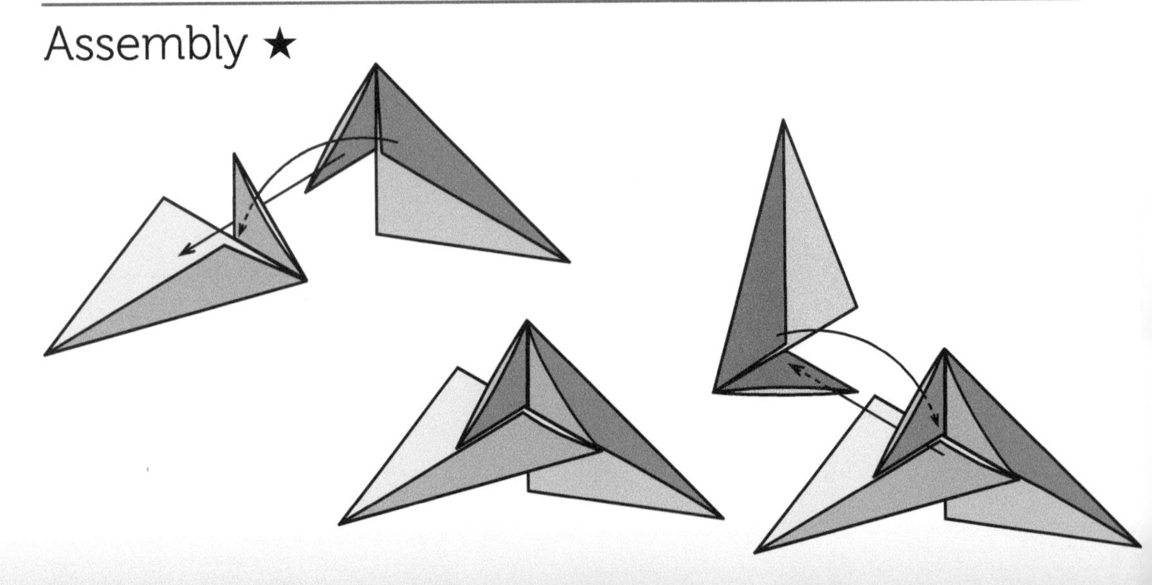

1 Tuck the flap of the first unit into the pocket of the second unit.

2 Two units joined.

3 Add the third unit. Note that there are only four sets of pockets like this.

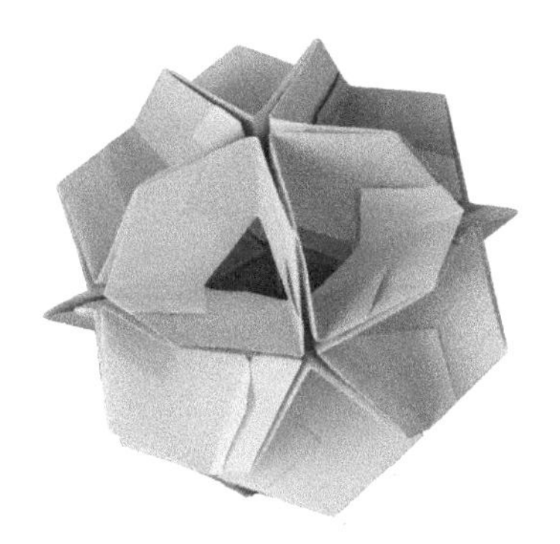

Star Ball

★★★
Use 30 double bronze rectangles that can be cut from five or six sheets of A4
Finished model ratio 1.6

Six double bronze rectangles can be cut from a sheet of A4 with a small amount of waste. You can use five or six colours. For five colours, each colour corresponds with the faces of a cube. For six colours, each colour forms a band.

Cutting Plan ★★

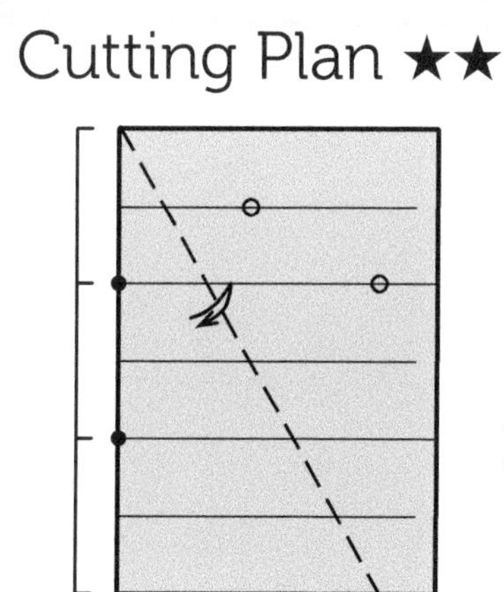

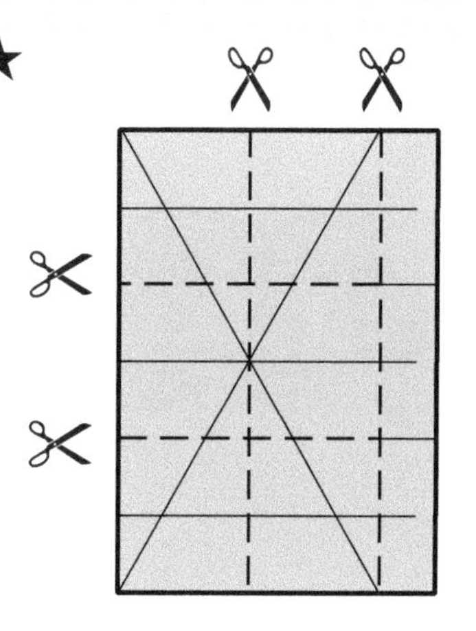

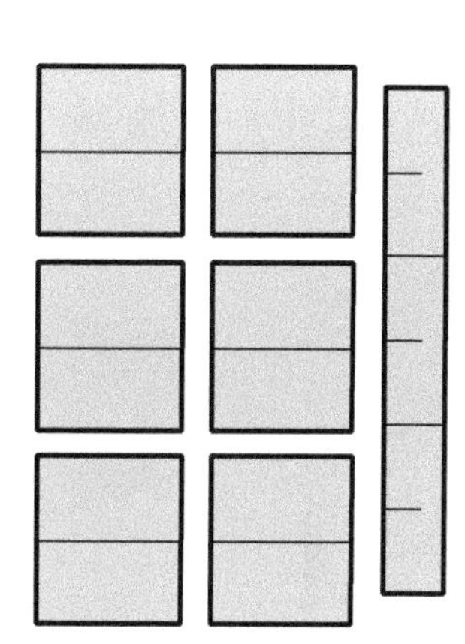

1 Divide the long edge into sixths. Fold 60°.

2 Cut the long strip first. Then cut the remaining rectangle into six.

3 Discard the long strip to leave six double bronze rectangles.

Module ★★

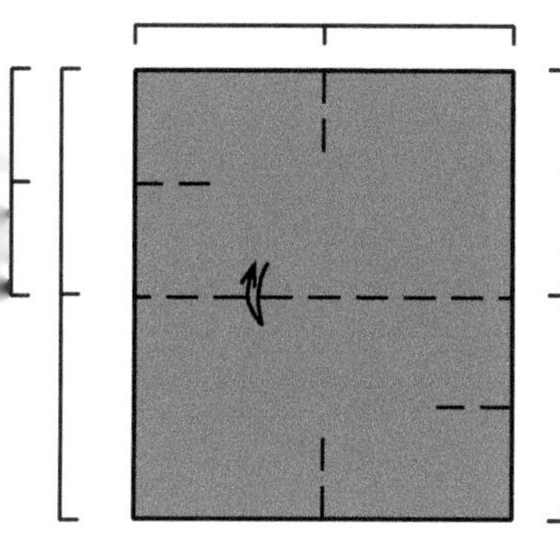

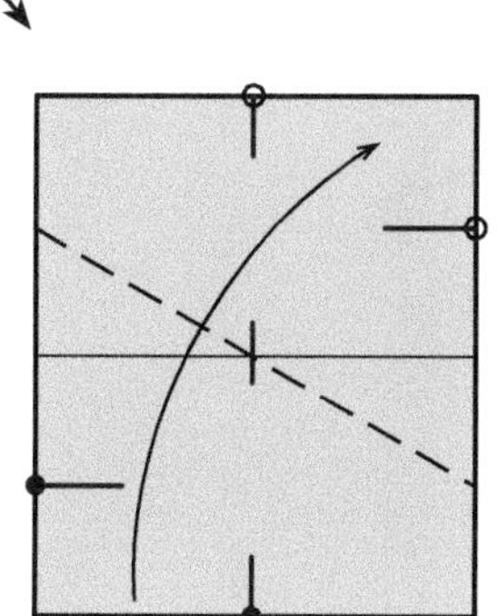

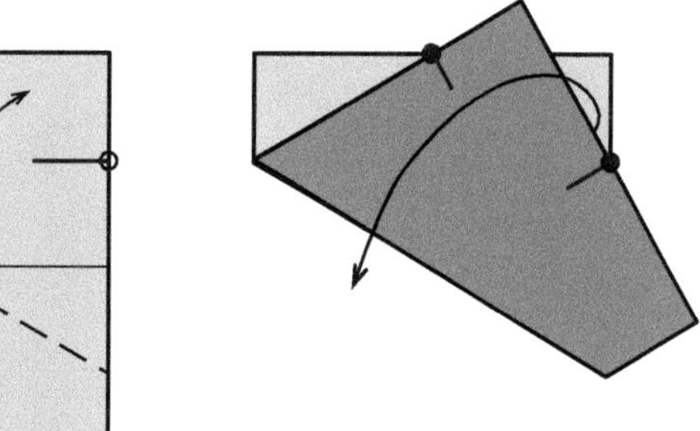

1 Pinch midpoints on the short edges and quarter points on the long edges. Turn over.

2 Fold quarter points to midpoints.

3 Unfold.

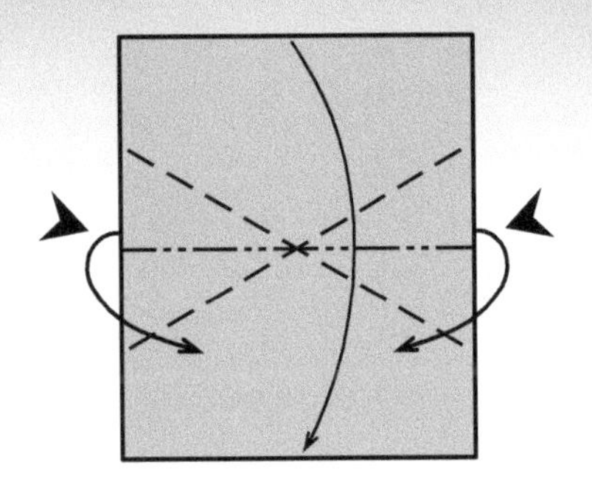

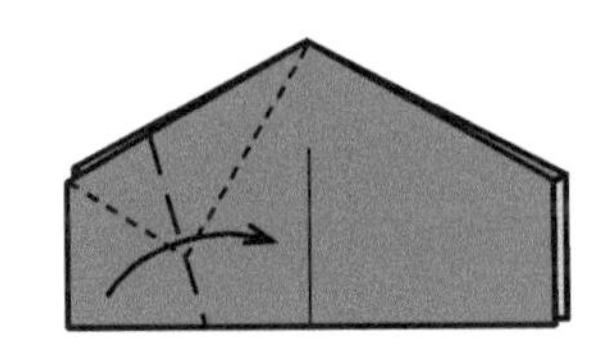

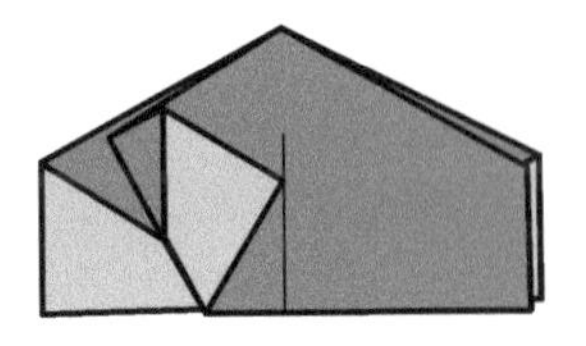

4 Fold in half, making two inside reverse folds, like a waterbomb base.

5 Put the corner to the central line, folding through the corner inside.

6 Repeat on the other three flaps.

Assembly ★★★

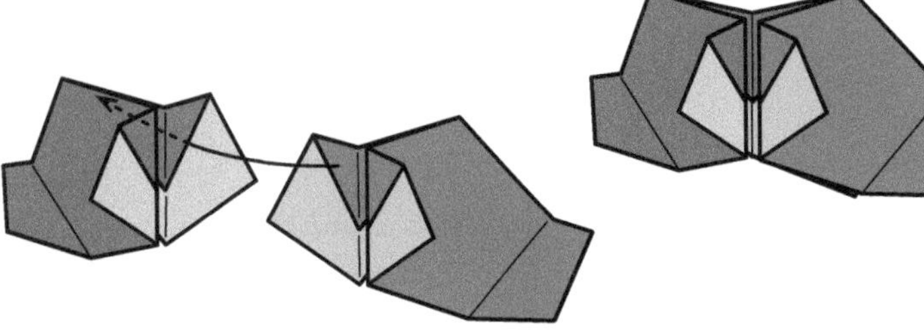

1 Insert the first unit into the second.

2 Join a group of five to make one star. Add five more units to make a pentagon. There are two main colour schemes.

Six Colours

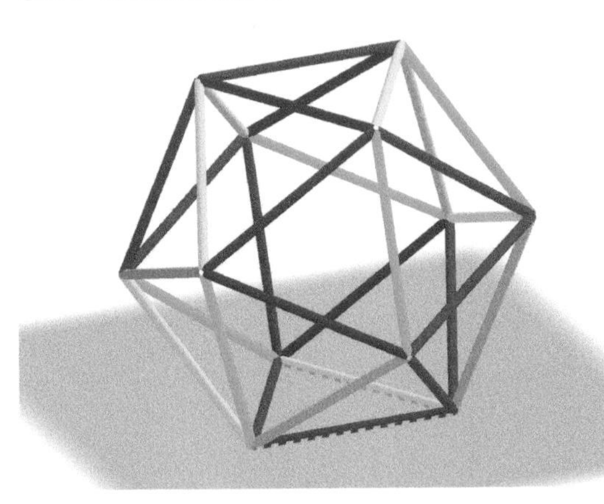

A regular icosahedron can be dissected into three objects: two pyramids with pentagonal bases and a pentagonal antiprism. Each colour forms a band on the wireframe antiprism: the units have rotational symmetry.

Five Colours

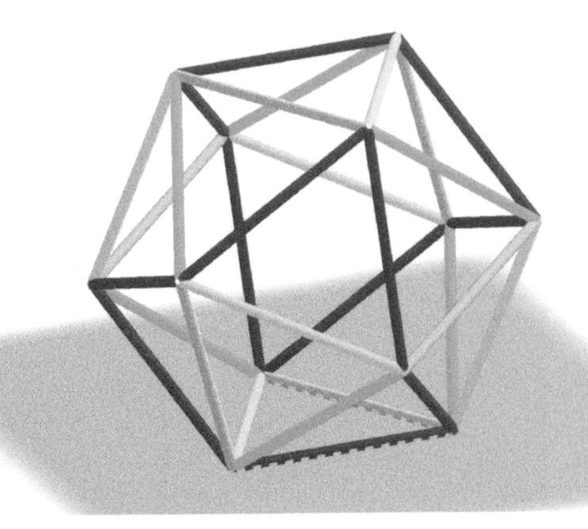

A regular icosahedron can be contained within a cube: each of six edges will lie on each face of a cube.

WXYZ

★★★
Use twelve squares
Finished model ratio 1.0

This was the result of experimenting with skeletal cuboctahedra, which can be seen as four intersecting regular hexagons: what happens if the hexagons are transformed into equilateral triangles? Use four colours so that each triangle is a different colour. Sizes from 75 mm are effective.

Module ★★★

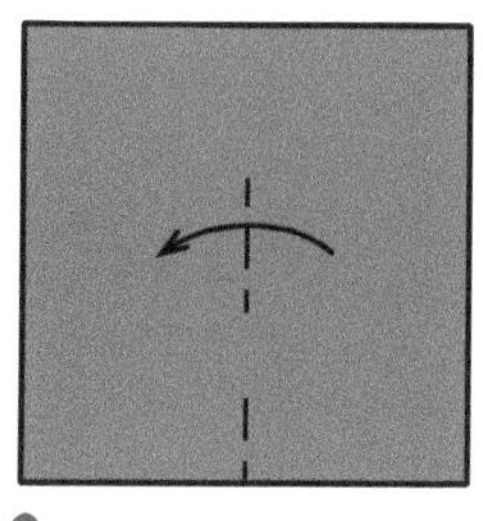

1 Pinch middle and bottom.

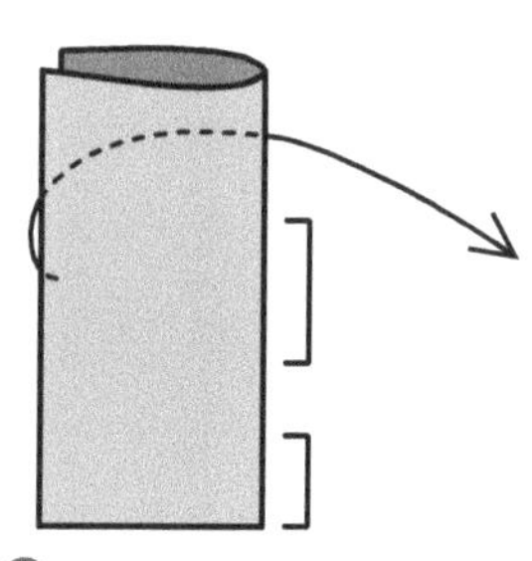

2 Unfold so that the white side is uppermost.

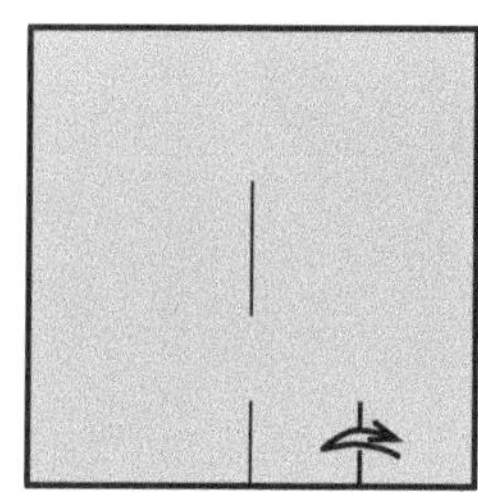

3 Pinch one quarter at bottom right.

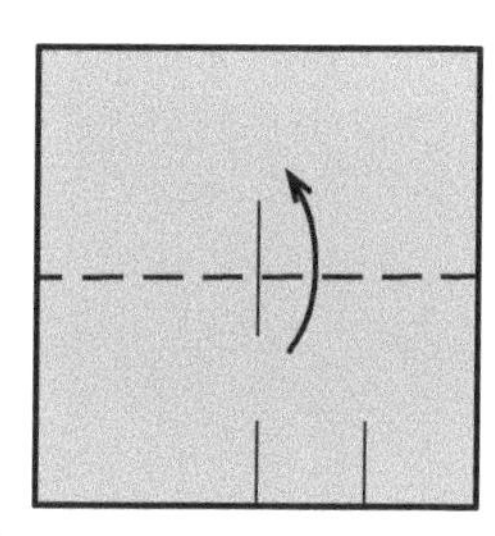

4 Valley in half upwards.

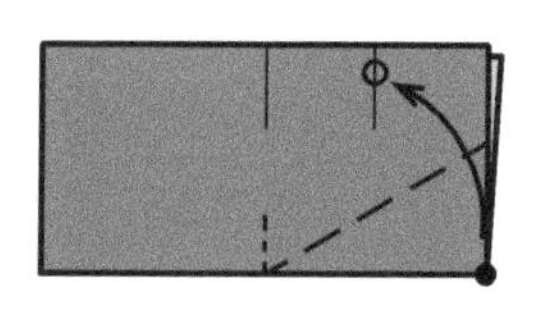

5 Use the mountain at the centre to help locate the fold.

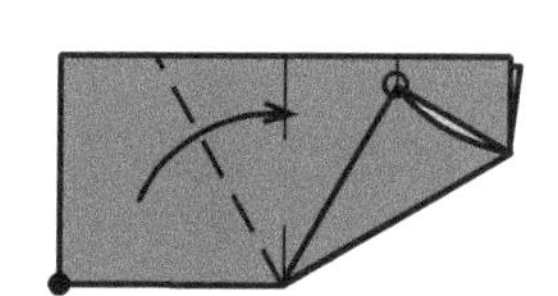

6 Fold the bottom left edge to meet the edge just folded.

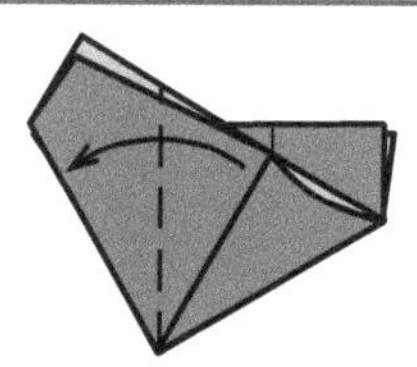

7 Fold the right edge just folded to meet the left edge.

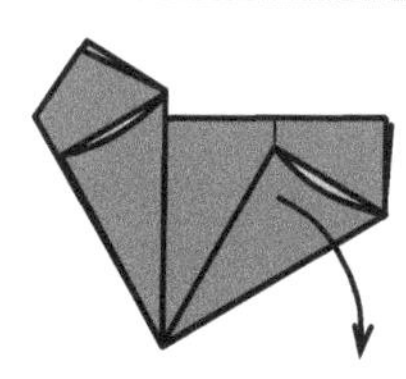

8 Unfold the right flap.

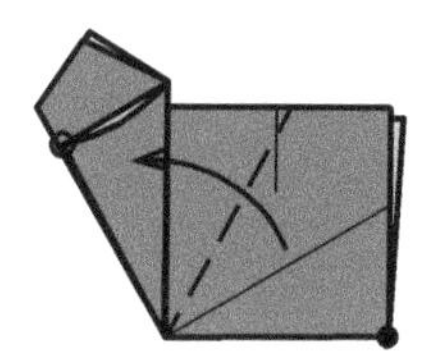

9 Fold the right corner to the left corner.

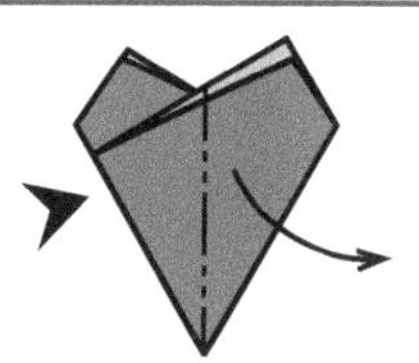

10 Squash the flap to the right.

11 Mountain fold the right flap behind.

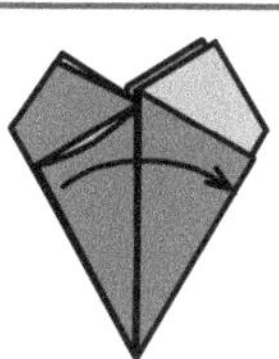

12 Fold the left flap to the left.

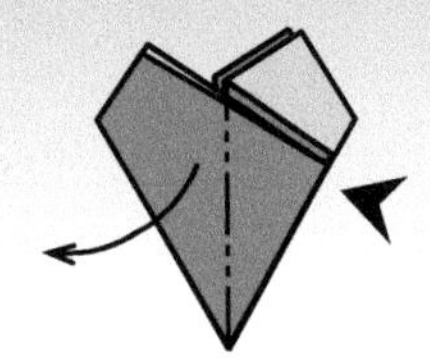

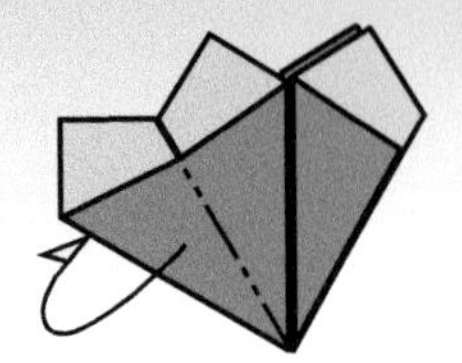

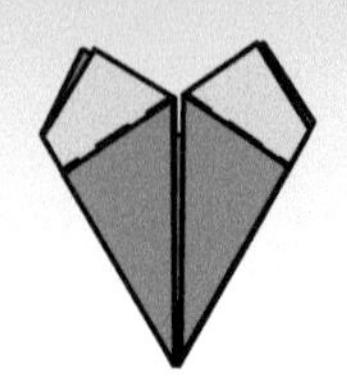

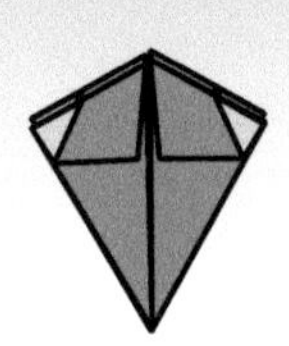

13 Squash the flap to the left.

14 Squash the flap to the right.

15 Fold down the white flaps against the coloured raw edges.

16 Stand the flaps at right angles.

Assembly ★★★

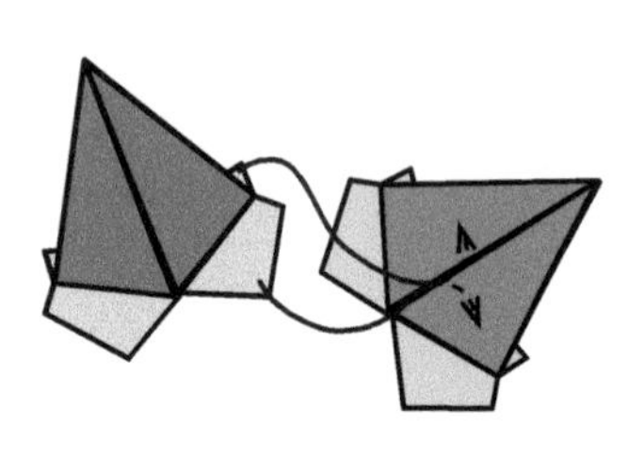

1 Each unit has a pair of pockets on each side. Tuck two flaps of the second unit into one pair of pockets.

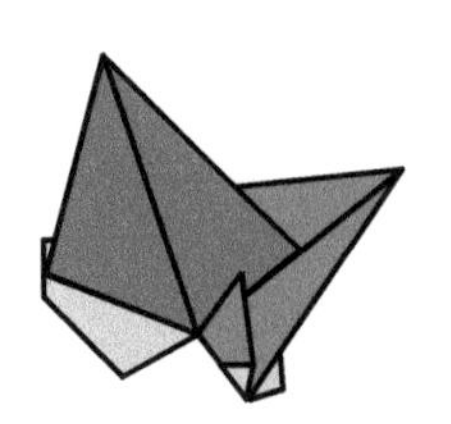

2 Two units joined. Add two more so that you have a ring of four units, each a different colour. Add more units to extend each plane of colour so that you have four triangles, each a different colour.

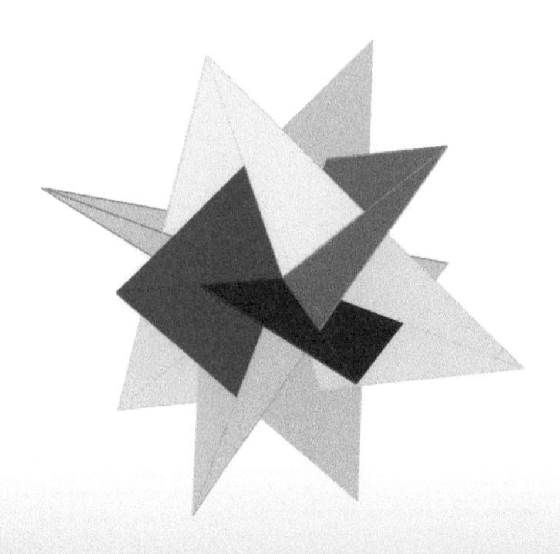

Blintz
Icosidodecahedron

★★★
Use thirty 75 mm squares or larger
Finished model ratio 2.0

Using six colours will help with assembling the units. Each of the six colours forms a flat five-pointed star. Remember to follow the pattern: there are only groups of five and groups of three. You might need to use a toothpick or file to open pockets and tuck flaps in near the end of assembly.

Module ★★

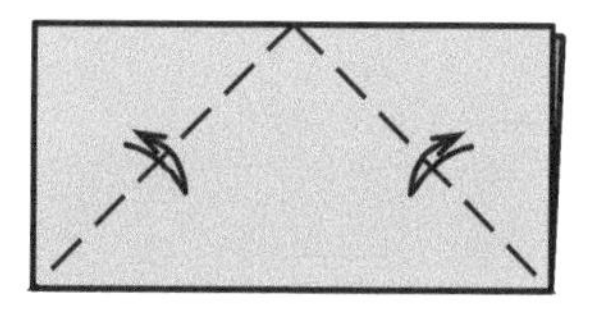

1 Start with a square folded in half. Fold the corners to the midpoint of the bottom edge.

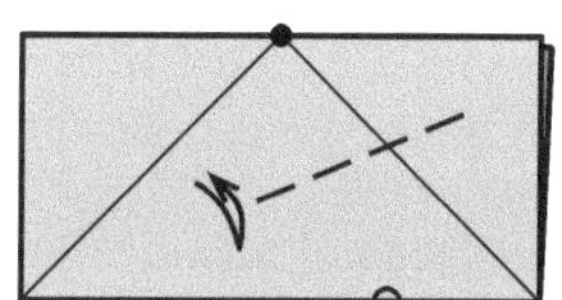

2 Bisect the left 45° angle and pinch where it meets the right crease.

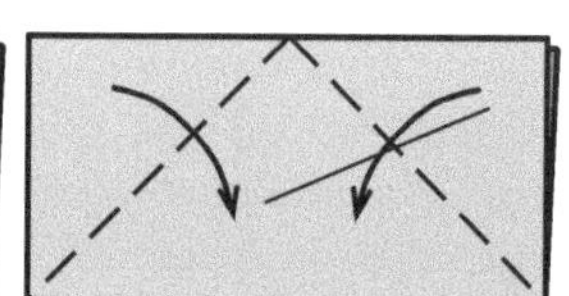

3 Refold the corners.

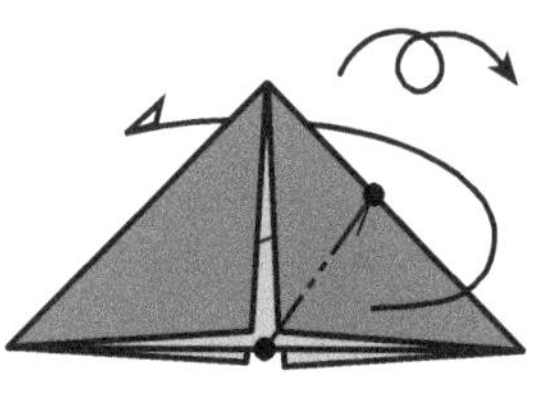

4 Fold the right flap behind using the location point from step 3 and the midpoint. Turn over.

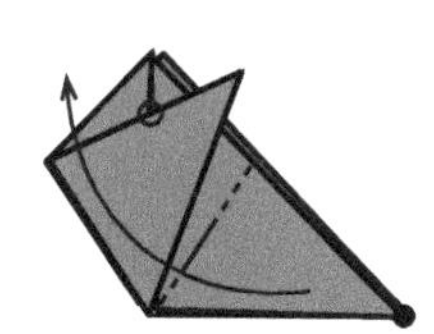

5 Fold the left flap to meet the location point.

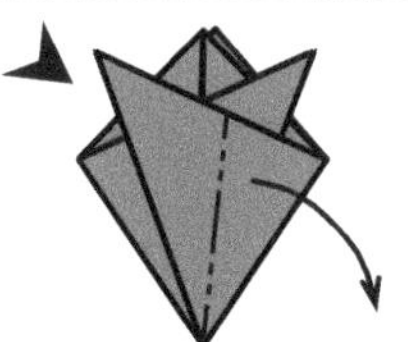

6 Squash fold the right flap.

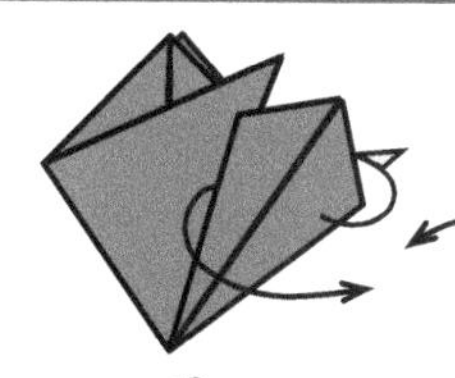

7 Flip the new flap behind. Squash the left flap.

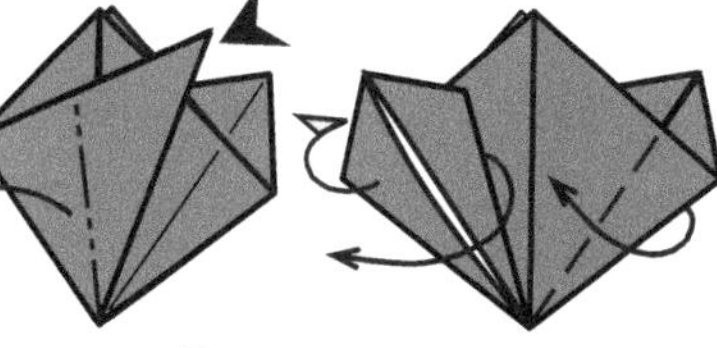

8 Make the new flaps stand out.

Assembly ★★★★

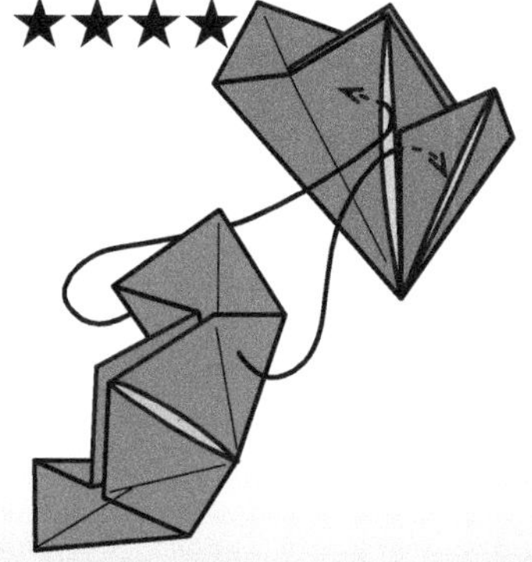

1 Insert the flaps of the first unit into the pocket of the second unit.

2 Continue adding units. Build a ring of five, then continue the pattern of colours.

Chapter 12

Notes for Teachers

Origami is a simple and accessible activity for learning mathematics. It allows learners to explore important mathematical topics and solve problems. Paper is a relatively cheap practical resource for classroom use. Any practical activity can enhance learning by making it more memorable.

Folding paper is an inherently mathematical activity: folding a point onto another creates a perpendicular bisector, folding repeatedly in half easily creates fractions of quarters and eighths. However, folding paper does not automatically teach mathematical concepts as much as walking does not necessarily teach you biomechanics.

This chapter is aimed at primary and secondary school mathematics teachers. However, it may be valuable for anyone interested in using origami for teaching and learning. First, common folding problems are described and advice given on how to avoid them. The next section describes tasks that can be used with almost any origami model. Then opportunities for learning some specific mathematical topics are given.

12.1 Common folding problems and how to avoid them

These problems are not restricted to children, they arise with all learners including teachers and university students. Suggestions are given for basic folding techniques that, if absent, prevent learners from engaging in *mathematical* activity.

Soft folding Most geometrical folds require sharp creases made with a finger or thumbnail.

Inaccurate folding caused by failing to line up edges precisely. For example, when folding a diagonal of a square, some learners try to make a fold through two corners. However it is much easier to join opposite corners together and make the crease by sweeping the paper flat away from the joined corners. See figure 12.1 (left) and figure 12.2.

Difficulty manipulating paper Beginners benefit from folding on a hard surface because of the extra assistance in supporting the paper. Positioning the paper is important. For example, to fold a square into a 2 by 1 rectangle,

DOI: 10.1201/9781003184492-12

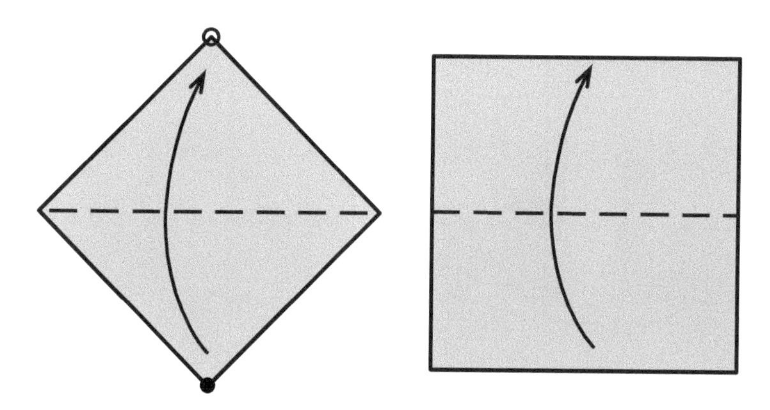

Figure 12.1: Left: fold the diagonal of a square by joining opposite corners. Right: "Book fold" with the "spine" closest to the body, not farthest away, to make folding the paper easier.

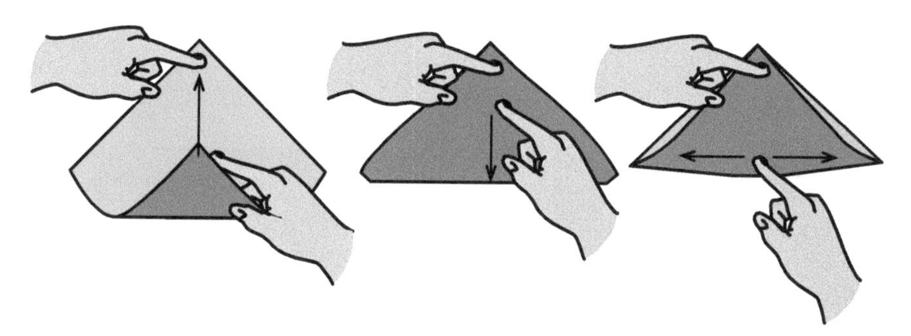

Figure 12.2: Fold the diagonal of a square on a hard surface by sweeping the paper, picking up the lower corner and placing it on the upper corner, sweeping downwards and creasing from the centre outwards.

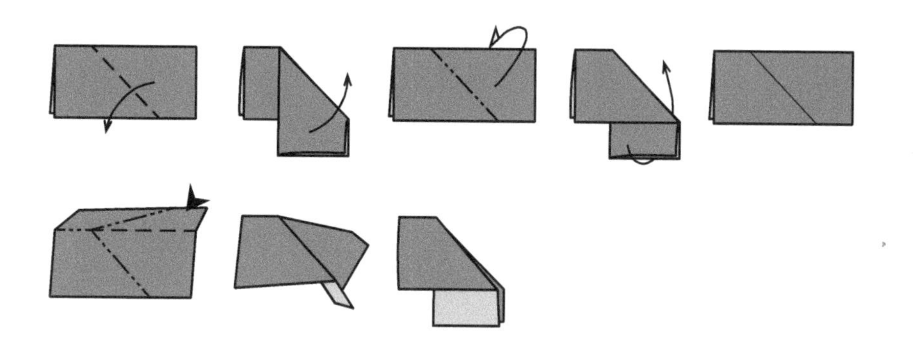

Figure 12.3: Precrease a reverse fold so that it is easier to make.

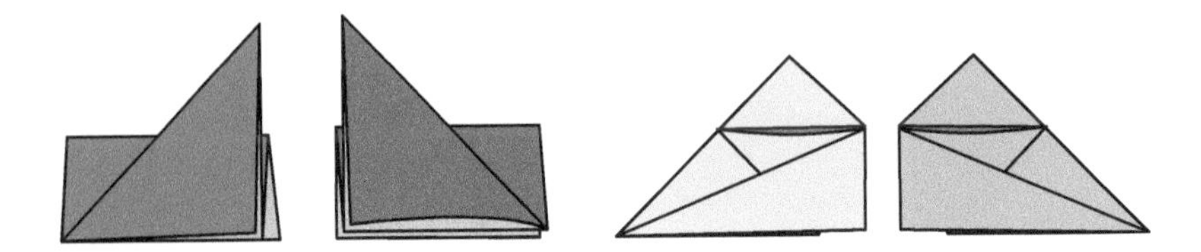

Figure 12.4: Left- and right-handed modules: use only one type in one model. Left- and right-handed modules for *Octagram Wreath* (p. 62) and *Octagram Paper Cup - Kite Wheel* (p. 63).

Figure 12.5: A poster of instructions for folding the traditional *Shuriken*.

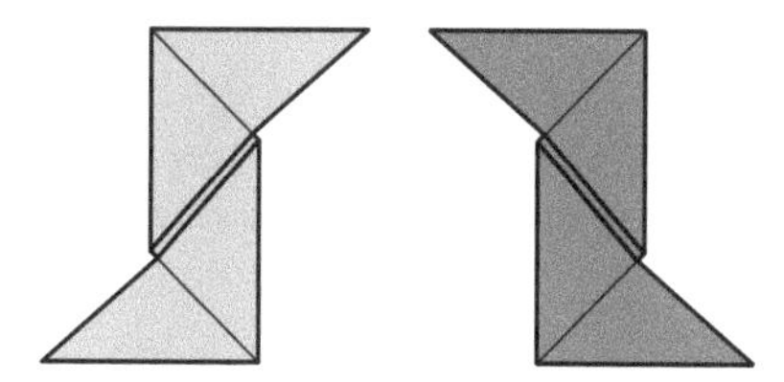

Figure 12.6: Left- and right-handed modules: a combination of both types is required in a model (traditional *Shuriken*).

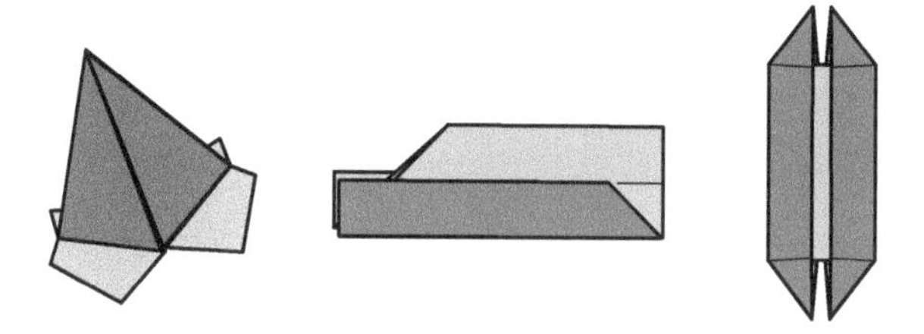

Figure 12.7: Achiral modules are symmetric so that they can be superimposed on their mirror images. Left to right: *WXYZ* (p. 109), *Pinwheel Square Slider* (p. 95) and *Square Pointer Slider* (p. 94).

it's easier to "book fold" with the "spine" closest to the body, not furthest away (figure 12.1, right).

Halving Whilst many learners can halve a side, halving a fold or an angle often presents particular challenges. For bisecting an angle, concentrate on lining up the edges or crease lines.

Reverse folds Beginners often find it difficult to do this, partly because their original crease is too soft. Folding backwards and forwards will help to reinforce the crease and achieve success (figure 12.3).

Mirror image modules Many modular models need to be folded in a consistent way to ensure they will fit together, e.g. *Octagram Ring, Square from Half Square* (p. 62) and *Octagram Paper Cup - Kite Wheel* (p. 63) (figure 12.4).

Learners commonly end up with a mixed set of right- and left-handed units that won't fit together. The traditional *Shuriken* (figure 12.5) star is invaluable for exploring this difficulty as it is made of two units that *are* mirror images (figure 12.6).

Of course some modules are not chiral, e.g. *WXYZ* (p. 109), *Square Pointer Slider* (p. 94) and *Intersecting Squares* (p. 89) (figure 12.7).

Visualisation and understanding instructions Encourage learners to adopt "help your neighbour" and "if you can do it, help someone else" so that the responsibility for success is shared throughout the group.

Some problems can be compounded by poor choice of paper, e.g. too small, too big, too soft, too thick, etc. When adopting Bernd Wollring's reverse engineering approach it is helpful to use paper that is identical to that used in the sample models. A larger piece of paper needs to be used for demonstrating, but it needs to be the same type as that used by the learners, e.g. only use duo paper if the students are using duo paper (and vice versa). Even if the model will work with any sized rectangle, (e.g. the traditional magazine box) ensure the paper for demonstration is similar to that used by the learners; this makes A paper particularly valuable.

12.2 General tasks

You can use these tasks with almost any origami model. A few examples are given but you will be able to come up with many more examples relevant to you.

12.2.1 Use mathematical vocabulary

Each step of a model is an opportunity to use mathematical vocabulary, both in explaining steps and describing the shapes made:

- What shape is this? [equilateral triangle, rectangle including square (and oblong), parallelogram, kite, irregular pentagon, regular octagon, etc.]
- Why? How do you know? [all lengths are the same and all angles are the same, etc.]

12.2.2 Identify the mathematical properties of shapes

You can ask at each step of the model about, for example, angle, length and symmetry. You can also unfold the paper and ask similar questions. For example,

- What kind of angle is this? [obtuse, right, acute or reflex angle]
- Estimate the angle? [about 30°, between 90° and 130°]
- What is the relationship between the edge length of a square and the length of the diagonal? [the diagonal is longer than the edge, the diagonal is the edge length multiplied by $\sqrt{2}$]
- What happens when you fold one point onto another? [You fold the perpendicular bisector of the line segment joining the points]

12.2.3 Use reverse engineering for teaching and learning

Challenge learners in groups to make an existing origami object. Each group has *two* examples of a folded object. Learners are advised to dismantle just one object and figure out how it is made. As the objects are modular (i.e. made of more than one piece) once they have decided how one unit is made they can then work together to produce the units they need. By having one intact model they can figure out how to reconstruct the model.

Simple models work well, e.g. *Star 8 Octad* (p. 68).

Each of these tasks can be followed up by asking learners to communicate their findings by preparing posters. One aim of these posters is to communicate to other groups of learners how to make the same object. In order to emphasise visual and geometric understanding, the constraint of using as few words as possible can be given. The use of *step folds* allows learners who cannot draw neatly with pencils and rulers to produce quality work (figure 12.5).

12.2.4 Practical work with fractions of length, angle and area

Folding paper is a natural context for practical work with fractions of length, angle and area. Halving is the easiest fraction to explore. You can halve the area of a rectangle by folding two congruent halves: either by joining opposite edges or by folding a diagonal.

Repeatedly halving gives quarters, eighths, sixteenths, etc. Folding other fractions without a marked straight edge is possible [Hatori] [Guy 75]. Figure 4.9 showed eight methods for dividing a square into thirds. How can you prove each method works? What are the advantages and disadvantages of each method? Which methods can be generalised to oblongs and other divisions, e.g. fifths and sevenths? Symbols and Procedures (p. 43) showed Fujimoto's iterative method for fifths.

Bisecting angles is similar to bisecting lengths and area.

12.2.5 Scale a design for a specific size

Practical experience is valuable: doubling the paper length quadruples the area. For 3D models, the volume increases by a factor of eight.

Sometimes a harder problem is to make the same model smaller and fit inside the original model (figure 12.8).

12.2.6 Explore the properties of A series paper

The silver rectangle described A series paper, a form of silver rectangle. Here are some mathematical activities.

One of the most common sizes of paper is A4. 'A' paper has a very special property: when folded in half along the short mirror line the resulting rectangle is similar and called A5; repeating gives A6, A7, etc. (figure 12.9 and figure 12.10). How many pieces of A10 paper can be made from a single sheet of A4 paper?

Figure 12.11 shows how to determine the proportions of a silver rectangle. The whole rectangle has proportion $1 : x$ (ratio of short to long sides) and the smaller rectangle $\frac{x}{2} : 1$.

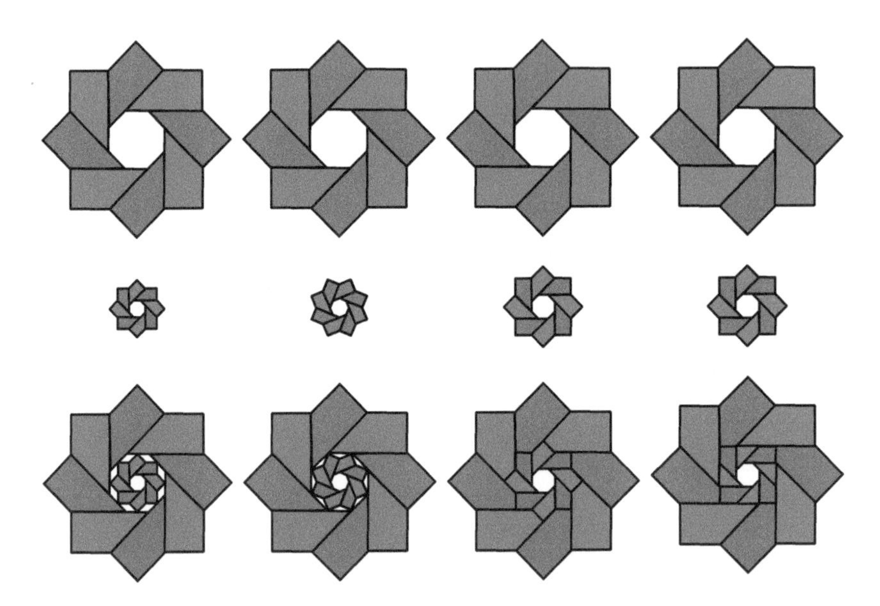

Figure 12.8: Fit a smaller model inside a larger model: what sizes of paper are needed? There are several choices about how the smaller model fits the larger model. From left to right: touching midpoints, touching vertices and aligning edges. The last choice tucks the red tips of the smaller model under the larger model.

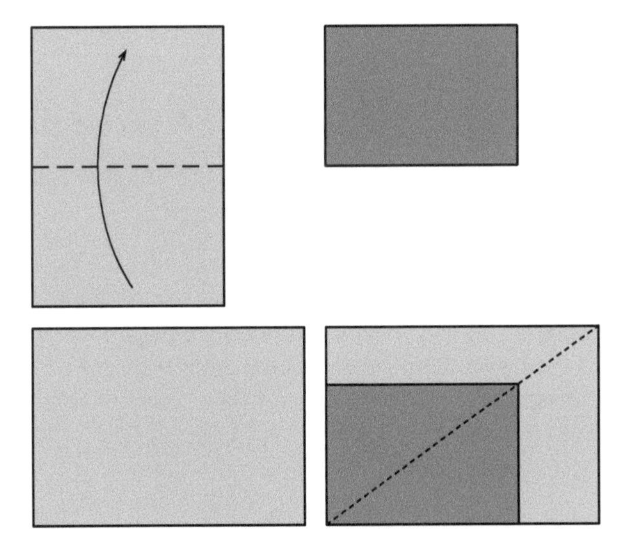

Figure 12.9: Take two sheets of A4. Fold one sheet in half and align it with the unfolded sheet to show that the two rectangles are of the same proportion.

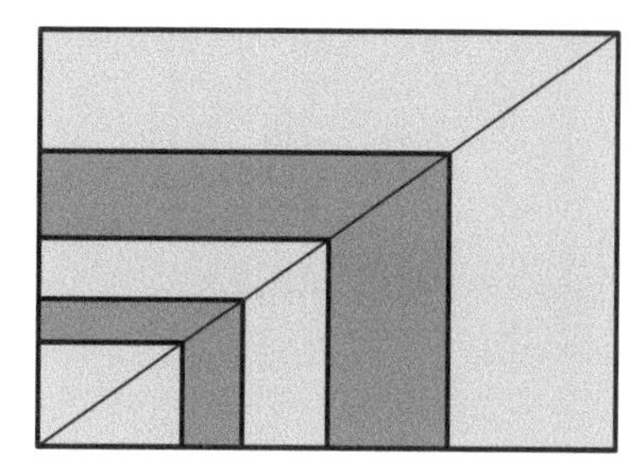

Figure 12.10: Extension of the previous method: after folding in half, cut in half and repeat the process.

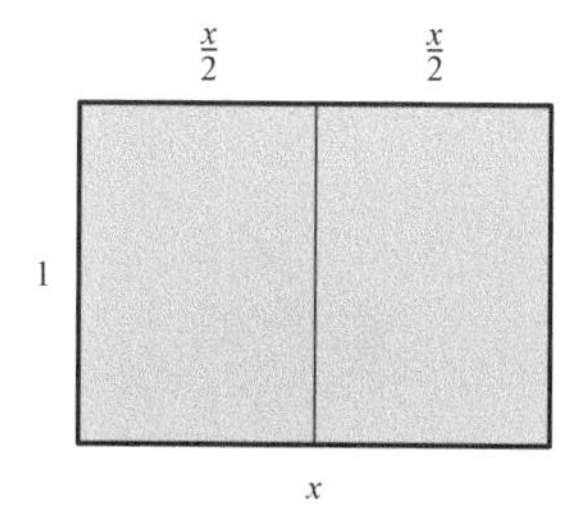

Figure 12.11: Determining the proportion of A4 paper.

The proportions are equal so $\frac{1}{x} = \frac{\frac{x}{2}}{1}$

or simply $\frac{1}{x} = \frac{x}{2}$

Multiply both sides by x $1 = \frac{x^2}{2}$

and then multiply both sides by 2 $2 = x^2$

Take the positive square root of both sides $x = \sqrt{2}$

This can be confirmed empirically with measurement and calculation. A bronze rectangle can be dissected into three similar rectangles: what are its proportions? How could you generalise this? What about generalising to triangles (figure 12.12)? What's special for $4, 9, 16, ..., n^2$ smaller similar triangles?

12.2.7 Generalise a design

You can generalise a model in several different ways:

- Number of sides

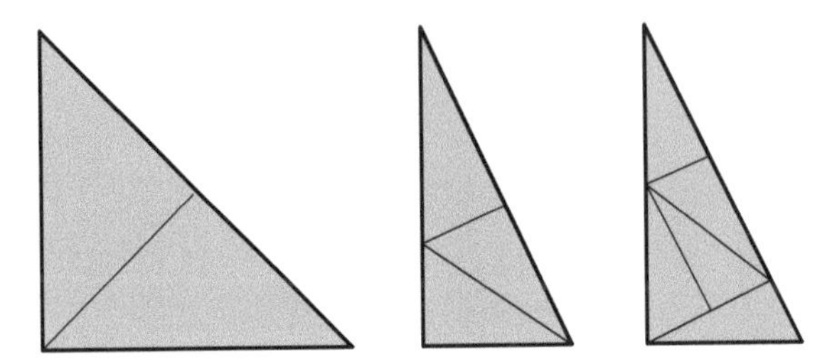

Figure 12.12: Triangles that dissect into two, three and five smaller similar triangles.

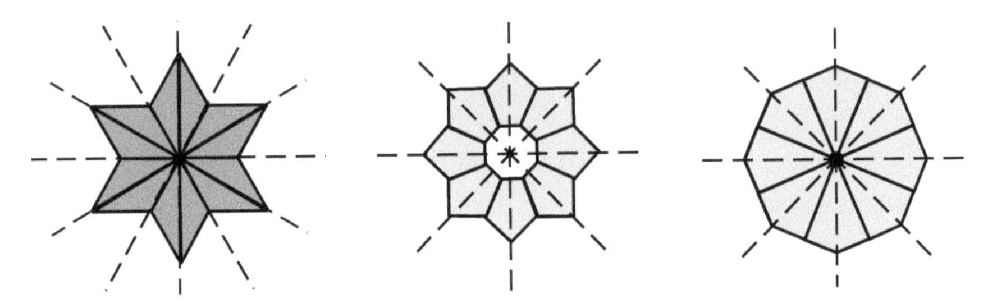

Figure 12.13: *Star 6* (p. 75) has six mirror lines and rotational symmetry of order 6. *Octagram Paper Cup - Kite Wheel* (p. 63) and *Quilt Star* (p. 90) have four mirror lines and rotational symmetry of order 8.

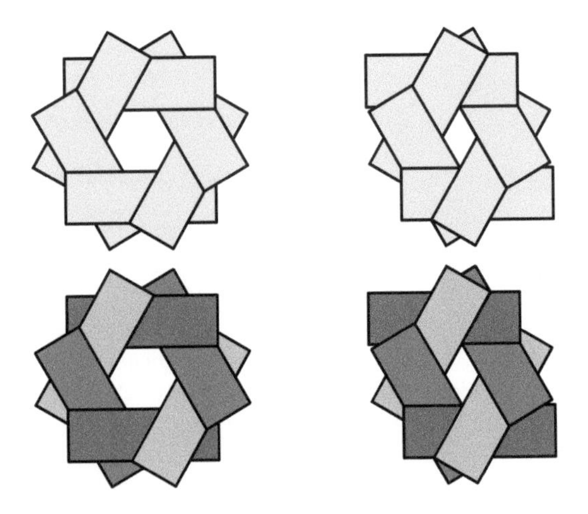

Figure 12.14: *Slider N 6* (p. 102) has rotational symmetry of order 6 (top left) and order 2 (top right and bottom row) in these configurations.

Figure 12.15: *Square Pointer Slider* (p. 94) has rotational symmetry of order 4, 2, 4 and 2 (top row) and order 2 (bottom row) in these configurations.

- Different starting proportions
- Different location points, e.g. shorter or wider
- Different angle

12.2.8 Identify the symmetries of a model

The most common symmetries are reflective (figure 12.13) and rotational (figure 12.14 and figure 12.15). Most origami stars and rings have rotational symmetry without mirror symmetry as the units are joined in one circular direction. Note that the symmetries depend on what you take into consideration: do you include colour and the interior details? For example, the outline may have reflective symmetry, but the interior does not.

For sliders, identify the symmetries as a model changes shape. Can you slide the shape so that it has no rotational symmetry?

12.2.9 Identify the transformations as a model changes shape

Most of the sliders in this book work by rotating or translating parts of the model. How could these be defined? Emphasise the information needed to fully specify the transformation:

2D Transformations

Translation with vector $\binom{\Delta x}{\Delta y}$

Rotation with angle r° and centre of rotation (x, y). Remember that a positive angle is an anticlockwise turn, by convention.

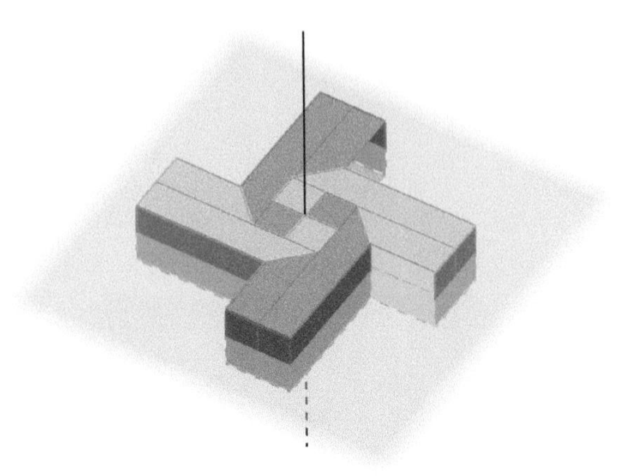

Figure 12.16: *Pinwheel Square Slider* (p. 95) has one mirror plane and one non-trivial axis of rotational symmetry.

Reflection with mirror line $y = mx + c$

Enlargement with scale factor r and centre of enlargement (x, y). Note that a scale factor between 0 and 1 gives a smaller result, and a negative scale factor is the same as a positive scale factor but with a rotation of 180° about the centre of enlargement.

Glide reflection with axis of glide reflection $y = mx + c$

Note that a model using two or more colours is likely to have a different order of symmetry to a model made of fewer colours (figure 12.14 and figure 12.15).

3D Transformations

The centre of rotation becomes an axis of rotation in 3D. Mirror lines become mirror planes in 3D. (figure 12.16)

12.2.10 Find the relationships, if any, between perimeter, area and volume

What is the relationship between surface area and volume, if any? [Use a counterexample: inverting the vertex of a cube preserves surface area but reduces the volume.]

12.2.11 Make a model in dynamic geometry software

Making a model in dynamic geometry software like Geogebra uses mathematical concepts, terminology and definitions like coordinates, transformations and vectors in a practical and motivating context. It also can use computing concepts like variables, loops, conditional statements, Boolean logic and data representation of colour. Problem-solving skills and creativity can be developed.

Some models can be challenging, so choose simpler models first, e.g. *Star 4, Square From Silver Rectangle* (p. 46) or *Poly Diag Star 5* (p. 54). There are two kinds of subjects that are particularly rewarding. The first kind are subjects that can be animated like sliders (figure 12.17) and 3D objects that can be rotated (figure 12.18). The second kind are models that can be generalised, e.g. once *Poly Diag Star 5* (p. 54) is generalised, it is easy to make *Poly Diag Star 6* (p. 53) and *Poly Diag Star 8* (p. 55) (figure 12.19). In fact, it is possible to try out variations before making them in paper.

Figure 12.17: Animation frames of *Slider N 8* (p. 101).

Figure 12.18: Animation frames of *WXYZ* (p. 109).

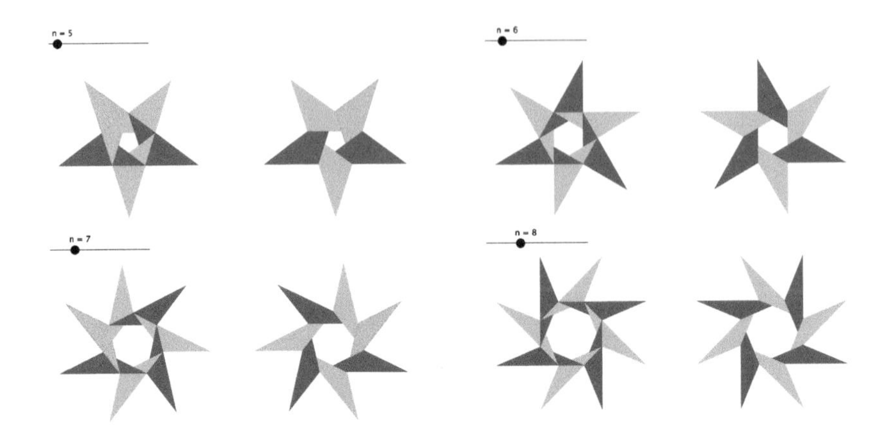

Figure 12.19: Top left shows the front and reverse sides of *Poly Diag Star 5* (p. 54) modelled in dynamic geometry software. Changing the value of n generates the other designs.

12.3 Specific mathematical topics

12.3.1 Folding 60° angles

Several models use 60° geometry such as *Star 6, Bronze Rectangle from Square* (p. 77), *Star 3, Square From Silver Rectangle* (p. 73) and *WXYZ* (p. 109). The method in figure 12.20 can easily be proved using symmetry. Point X lies on the perpendicular bisector of L and R and is equidistant from L and R. The distance from L to X is the same as that from L to R. The distance from R to X is the same as that from R to L. Triangle XLR is equilateral so the angles in the triangle are 60°.

Figure 12.21 shows a similar proof but draws an analogy with the method for constructing equilateral triangles with a pair of compasses.

The method of folding 60° is a special case of trisecting an angle. Figure 12.22 shows how to trisect an acute angle by folding: this is not possible using an unmarked straight edge and a pair of compasses. Figure 12.23 shows how to add some extra lines to enable a proof.

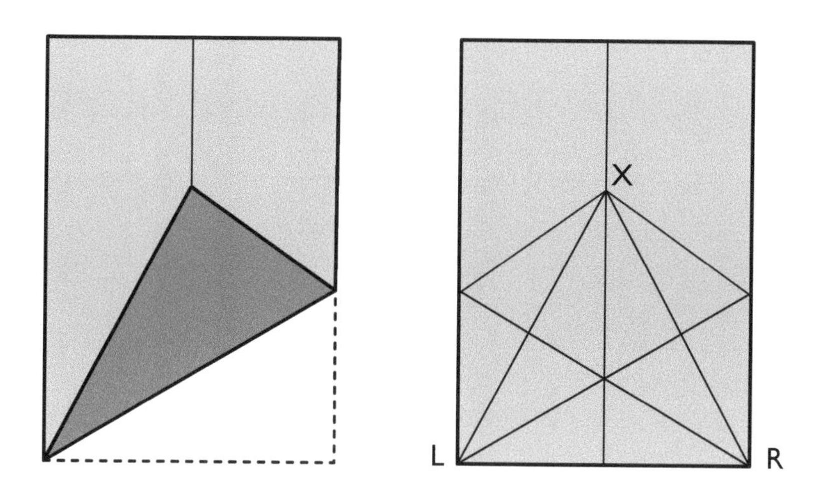

Figure 12.20: Trisecting a right angle: proof by symmetry.

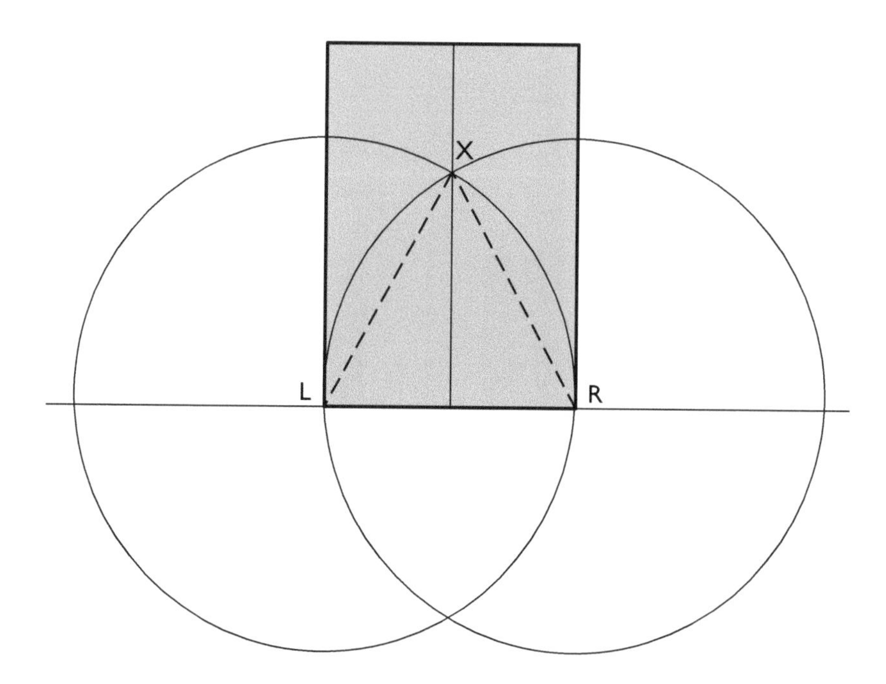

Figure 12.21: Trisecting a right angle: proof by analogy with the construction using a pair of compasses.

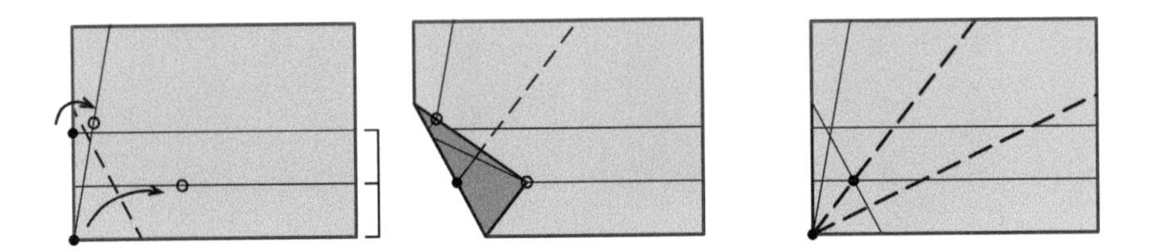

Figure 12.22: Abe method for trisecting an acute angle by folding.

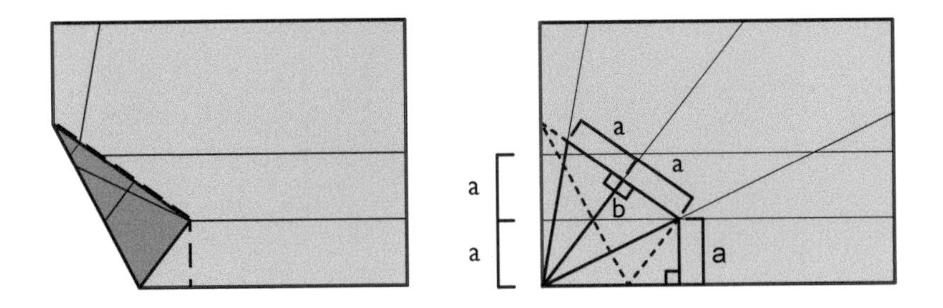

Figure 12.23: Construction to prove Abe trisection.

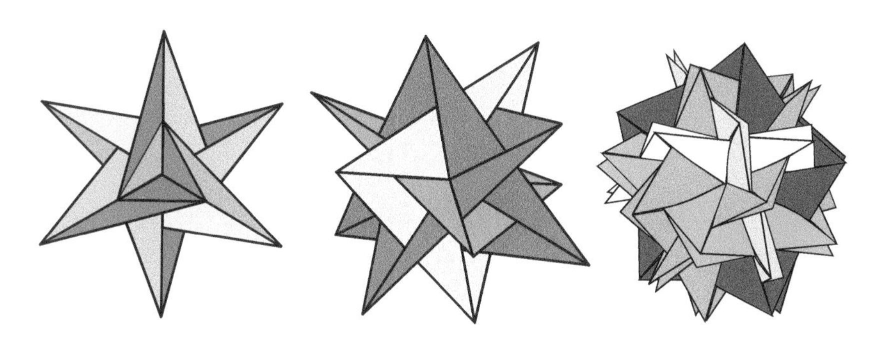

Figure 12.24: Nolids: *XYZ Rhombic* (p. 106), *WXYZ* (p. 109) and *Blintz Icosidodecahedron* (p. 111) are based on the skeletal regular octahedron, cuboctahedron and icosidodecahredron, respectively.

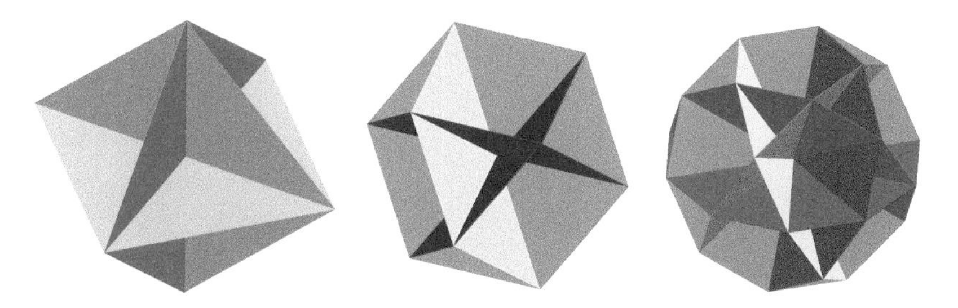

Figure 12.25: The skeletal regular octahedron, cuboctahedron and icosidodecahredron.

12.3.2 Nolids

Figure 12.24 shows *XYZ Rhombic* (p. 106), *WXYZ* (p. 109) and *Blintz Icosidodecahedron* (p. 111). They are related to nolids (solids of zero volume), also known as skeletal polyhedra (figure 12.25).

The skeletal regular octahedron can be modelled in a couple of different ways

- three squares that intersect each other
- replace the faces of a regular octahedron with polygons that join the edges with the centre

How could these be applied to other polyhedra?

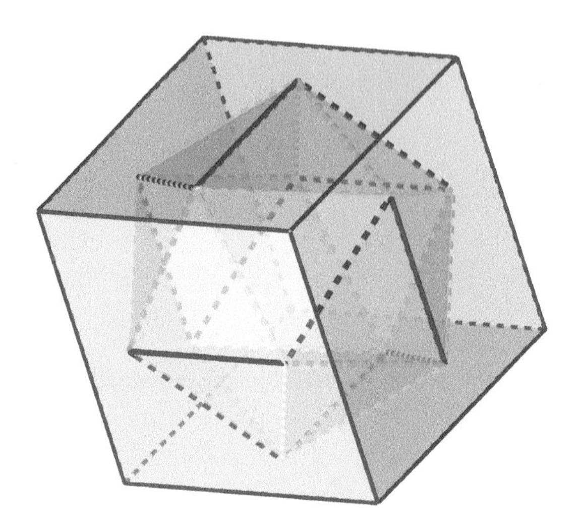

Figure 12.26: The red edges of the icosahedron lie on the faces of a cube.

Figure 12.27: The four red faces of the regular icosahedron are separated by grey faces.

12.3.3 The regular icosahedron and other regular polyhedra

The instructions for the five colour version of *Star Ball* (p. 107) state that edges of the same colour lie on the faces of a cube. Figure 12.26 shows that the red edges of the icosahedron lie on a cube. Colouring faces so that each face is surrounded by faces of a different colour is a classic problem. The icosahedron in figure 12.27 has four red faces that hint at one solution.

Other fascinating relations exist between the pdfregular polyhedra. For example, compound polyhedra exist like the stella octangular which is a compound of two tetrahedra and fits inside a cube. The books cited in Mathematics and geometry, p. 130, have more like this.

Chapter 13

Further Reading

Four broad categories of further reading are given here. Not all books are currently in print so you might need to look in second-hand bookshops or libraries. Try searching for websites if they become unavailable or change address.

13.1 Origami

Many origami books contain a few stars and rings. For example, Nakano has some simple rings that need some glue [Nakano 85]. Some young people created their rings in Russia [Afonkin and Hull 98]. Not only is Robert Neale responsible for the classic slider *Pinwheel-Ring-Pinwheel* (also known as *Magic Star*), he has also created many other elegant models [Neale and Hull 94].

Two books by Vincente Palacios feature more than the usual number of flat modulars [Palacios 00] [Palacios 03]. The earlier and shorter book has many flat modulars by Francisco Javier Caboblanco, as well as some by Jose Meeusen. The other book has some models by the late Jeff Beynon and others. Beynon compiled four books of his own designs in his distinctive graphic style [Beynon 89] [Beynon 90] [Beynon 91] [Beynon 93]. They all have at least a few flat modulars, and all are described in an abbreviated way with few steps but full text instructions. One of my earlier books has several sliders and a few more planar models [Lam 18].

A few origami books are dedicated to modular origami stars. Maria Sinayskaya's book has three simple flat stars and 17 3D works [Sinayskaya 16]. Tomoko Fuse compiled many flat modular origami stars and rings [Fuse 07]. The late Dave Petty also wrote a couple of dedicated books [Petty 94] [Petty 98]. The earlier book has a series of models exploring the traditional paper cup as a basis for origami rings and coasters. Carmen Sprung wrote a couple of books in German [Sprung 15a] [Sprung 15b]. Most of the stars are modular but some are made from one sheet, some of which are not rectangles. Mette Pederson is another self-published author [Pederson 10a] [Pederson 10b]. For stars made from a single sheet of paper, try the book by John Montroll [Montroll 14]. Paul Jackson created a system for making 3D stars using pre-cut units [Jackson 20].

Dave Petty compiled planar modular models in a publication by the British Origami Society [Petty 13]. *Intersecting Planes Gallery* shows *WXYZ* and other planar models [Mukerji 03]. The late Francis Ow created many sliders and woven stars [Ow 18a] [Ow 18b].

DOI: 10.1201/9781003184492-13

13.2 Mathematics and geometry

Two valuable books that have been a source of inspiration for many years are David Wells' *Curious and Interesting Geometry* [Wells 91], which gathers diverse geometrical theorems, oddities and facts, and Cundy and Rollett's classic *Mathematical Models*, which describes, analyses and organises the regular and semi-regular polyhedra [Cundy and Rollett 81].

Other books include a guide of instructions for making many complicated polyhedra [Wenninger 71], an accessible summary of polyhedra, their properties and the relationships between them [Cromwell 04] and a book aimed at the layperson that includes many intriguing pictures and their associated mathematics [Steinhaus 00].

13.3 Origami and mathematics

This field has grown quickly in the last few years but has much older roots: one book was originally written in 1893 [Sundara Row 66]. The mathematics of dividing lengths into equal parts is accessible to those with mathematics at secondary school level [Hatori] [Guy 75].

More advanced works include a book on how to design certain kinds of complex single sheet origami [Lang 11], research monographs [Demaine and O'Rourke 07] [Hull 20] and the proceedings of the International Meetings of Origami Science, Mathematics, and Education (OSME) [Lang 09] [Wang-Iverson et al. 11].

13.4 Origami and mathematics education

Several books use origami for teaching and learning mathematics. These can be divided by level (school or college/university) and type (descriptions of tasks or descriptions with specific pedagogic advice).

Project Origami has worksheets for undergraduates with pedagogical notes for teachers but several tasks could be adapted for secondary schools [Hull 13]. The OSME proceedings usually have several relevant papers [Lang 09] [Wang-Iverson et al. 11].

School-level books are typically at secondary school level (11 to 16 years of age). Two publications of tasks by the National Council of Teachers of Mathematics have a similar selection of classic material [Olson 75] [Johnson 99]. *Geometric Exercises in Paper Folding* is a little dated and hard to understand but is of historical interest [Sundara Row 66]. Kazuo Haga presents several elegant investigations in *Origamics* [Haga et al. 08].

Learning Mathematics with Origami is aimed at primary and secondary school teachers [Lam and Pope 16]. As well as giving practical advice on using origami for learning mathematics, it promotes opportunities for learners to think creatively, justify decisions, deepen understanding, take responsibility for their own learning and work collaboratively with others.

Bibliography

[Afonkin and Hull 98] Sergei Afonkin and Tom Hull. *Russian Origami*. St. Martin's Griffin, 1998.

[Beynon 89] Jeff Beynon. *Origami (BOS Booklet 27)*. British Origami Society, 1989.

[Beynon 90] Jeff Beynon. *More'igami (BOS Booklet 31)*. British Origami Society, 1990.

[Beynon 91] Jeff Beynon. *Jef Ori' 3 (BOS Booklet 37)*. British Origami Society, 1991.

[Beynon 93] Jeff Beynon. *Multiplication (BOS Booklet 44)*. British Origami Society, 1993.

[Biddle and Biddle 93] Steve Biddle and Megumi Biddle. *The New Origami*. Ebury Press, 1993.

[Cromwell 04] Peter Cromwell. *Shapes in Space: Convex Polyhedra with Regular Faces*. Association of Teachers of Mathematics, 2004.

[Cundy and Rollett 81] H. Martyn Cundy and A. R. Rollett. *Mathematical Models*, Third edition. Tarquin Publications, 1981.

[Demaine and O'Rourke 07] Erik D. Demaine and Joseph O'Rourke. *Geometric Folding Algorithms: Linkages, Origami, Polyhedra*. Cambridge University Press, 2007.

[Fuse 07] Tomoko Fuse. *Origami Rings and Wreaths; A Kaleidoscope of 28 Decorative Origami Creations*. Japan Publications Trading Company, 2007.

[Guy 75] Mick Guy. *Geometrical Division*. British Origami Society, 1975.

[Haga et al. 08] Kazuo Haga, Josefina Fonacier, and Masami Isoda. *Origamics: Mathematical Explorations Through Paper Folding*. World Scientific, 2008.

[Hatori] Koshiro Hatori. 'How to Divide the Side of Square Paper.' Available online (`http://www.origami.gr.jp/Archives/People/CAGE_/divide/index-e.html`).

[Hull 13] Thomas Hull. *Project Origami: Activities for Exploring Mathematics*, Second edition. A. K. Peters/CRC Press, 2013.

[Hull 20] Thomas C. Hull. *Origametry: Mathematical Methods in Paper Folding*. Cambridge University Press, 2020.

[Jackson 82] Paul Jackson. *Origami Christmas Tree Decorations.* British Origami Society, 1982.

[Jackson 20] Paul Jackson. *Superstars: Make a Galaxy of 3D Paper Stars.* Laurence King Publishing, 2020.

[Johnson 99] Donovan A. Johnson. *Paper Folding for the Mathematics Class.* National Council of Teachers of Mathematics, 1999.

[Kawamura 03] Miyuki Kawamura. 'Origami with Trigonometric Functions.' In *Origami 3: Third International Meeting of Origami Science, Math, and Education*, edited by Thomas Hull. A. K. Peters, 2003.

[Lam and Pope 16] Tung Ken Lam and Sue Pope. *Learning Mathematics with Origami.* Association of Teachers of Mathematics, 2016.

[Lam 18] Tung Ken Lam. *Action Modular Origami to Intrigue and Delight.* Tarquin, 2018.

[Lang 09] Robert J. Lang, editor. *Origami 4: Fourth International Meeting of Origami Science, Mathematics, and Education.* A. K. Peters, 2009.

[Lang 11] Robert J. Lang. *Origami Design Secrets: Mathematical Methods for an Ancient Art, Second Edition.* A. K. Peters/CRC Press, 2011.

[Mitchell 15] David Mitchell. 'David Mitchell's Origami Heaven - Origami Unfolded - Useful Rectangles.' Available online (`http://www.origamiheaven.com/usefulrectangles.htm`), 2015.

[Montroll 14] John Montroll. *Origami Stars.* Dover, 2014.

[Mukerji 03] Meenakshi Mukerji. 'Intersecting Planes Gallery.' Available online (`http://www.origamee.net/gallery/planars.html`), 2003.

[Nakano 85] Dokuohtei Nakano. *Easy Origami.* Beaver, 1985. Translated by Eric Kenneway.

[Neale and Hull 94] Robert Neale and Thomas Hull. *Origami, Plain and Simple.* St. Martin's Press, 1994.

[Nicholson et al. 98] Ben Nicholson, Jay Kappraff, and Saori Hisano. 'A Taxonomy of Ancient Geometry Based on the Hidden Pavements of Michelangelo's Lurentian Library.' In *Bridges: Mathematical Connections in Art, Music, and Science*, edited by Reza Sarhangi, pp. 255–272. Southwestern College, Winfield, Kansas: Bridges Conference, 1998. Available online at `http://archive.bridgesmathart.org/1998/bridges1998-255.html`.

[Olson 75] A. T. Olson. *Mathematics Through Paper Folding.* National Council of Teachers of Mathematics, 1975.

[Ow 18a] Francis Ow. 'Expandables.' Available online (`https://www.flickr.com/photos/61236172@N08/albums/72157662049397716`), 2018.

[Ow 18b] Francis Ow. 'Woven Stars.' Available online (`https://www.flickr.com/photos/61236172@N08/albums/72157627081139368`), 2018.

[Palacios 00] Vincente Palacios. *Origami for Beginners.* Dover Publications Inc., 2000.

[Palacios 03] Vincente Palacios. *Origami from around the World.* Dover Publications Inc., 2003.

[Pederson 10a] Mette Pederson. *Mette Units 4: Rings.* CreateSpace Independent Publishing Platform, 2010.

[Pederson 10b] Mette Pederson. *Mette Units 5 by Mette Pederson.* CreateSpace Independent Publishing Platform, 2010.

[Petty 94] David Petty. *Modular Construction and Twists (BOS Booklet 46).* British Origami Society, 1994.

[Petty 98] David Petty. *Origami Wreaths and Rings.* Aitoh, 1998.

[Petty 13] David Petty. *Planar Modular Origami.* British Origami Society, 2013.

[Plank 19] Jim Plank. 'The pentagon module (108 degrees).' Available online (`http://web.eecs.utk.edu/~jplank/plank/origami/penultimate/pentagon.html`), 2019.

[Sinayskaya 16] Maria Sinayskaya. *Zen Origami: 20 Modular Forms for Meditation and Calm: 400 sheets of origami paper in 10 unique designs included!* Race Point Publishing, 2016.

[Sprung 15a] Carmen Sprung. *Origami - 21 Sterne.* Carmen Sprung, 2015.

[Sprung 15b] Carmen Sprung. *Origami - 25 Sterne.* Carmen Sprung, 2015.

[Steinhaus 00] H. Steinhaus. *Mathematical Snapshots.* Dover Publications Inc., 2000.

[Sundara Row 66] T. Sundara Row. *Geometric Exercises in Paper Folding.* Dover Publications, 1966.

[Wang-Iverson et al. 11] Patsy Wang-Iverson, Robert J. Lang, and Mark Yim, editors. *Origami 5: Fifth International Meeting of Origami Science, Mathematics, and Education.* A. K. Peters/CRC Press, 2011.

[Wells 91] David G. Wells. *The Penguin Dictionary of Curious and Interesting Geometry.* Penguin, 1991.

[Wenninger 71] Magnus J. Wenninger. *Polyhedron Models.* Cambridge University Press, 1971.

Index

About the Author

Tung Ken Lam is an origami creator, author and qualified mathematics teacher. His best-known works are *WXYZ* (p. 109) and *Jitterbug*.

He is the author of *Action Modular Origami to Intrigue and Delight* (Tarquin, 2018), and coauthor of *Learning Mathematics with Origami* (Association of Teachers of Mathematics, 2016).

> His modular creations are fine examples of original and economical folding.
>
> –David Petty, author of *Origami 1-2-3* and *Origami A-B-C*

> It has been a long time since I have enjoyed folding so many projects from just one book. The diagrams are among the best I have seen for geometric origami, and the collection of projects is a satisfying variety. Highly recommended! Five out of five stars.
>
> –Michael LaFosse, cofounder of Origamido® Studio and author of over 50 origami books, on *Action Modular Origami to Intrigue and Delight*

He has taught and presented his origami work in France, Italy, Sweden, Japan and USA. He has run many events for teachers, learners and the general public. He accepts commissions for origami projects and creating bespoke origami to order. Clients have included Bletchley Park Trust, The Bodleian Library, University of Oxford, EuroStemCell (formerly the Scottish Centre for Regenerative Medicine) and Honda UK.

Visit `https://www.foldworks.net` to find out more.

Online Materials

`https://www.geogebra.org/u/tung+ken+lam` has Geogebra dynamic geometry files associated with this book. These resources are licensed using Creative Commons CC BY-SA 3.0, Attribution-ShareAlike 3.0.